# Progress in Improving Project Management at the Department of Energy

## 2003 Assessment

Committee for Oversight and Assessment of U.S. Department of Energy
Project Management

Board on Infrastructure and the Constructed Environment

Division on Engineering and Physical Sciences

### NATIONAL RESEARCH COUNCIL
*OF THE NATIONAL ACADEMIES*

THE NATIONAL ACADEMIES PRESS
WASHINGTON, D.C.
**www.nap.edu**

**THE NATIONAL ACADEMIES PRESS**   500 Fifth Street, N.W.   Washington, DC 20001

NOTICE: The project that is the subject of this report was approved by the Governing Board of the National Research Council, whose members are drawn from the councils of the National Academy of Sciences, the National Academy of Engineering, and the Institute of Medicine. The members of the committee responsible for the report were chosen for their special competences and with regard for appropriate balance.

This study was supported by Contract Number DEAM01-99PO8006 between the U.S. Department of Energy and the National Academy of Sciences. Any opinions, findings, conclusions, or recommendations expressed in this publication are those of the authors and do not necessarily reflect the views of the organizations or agencies that provided support for the project.

International Standard Book Number 0-309-09180-2 (book)
International Standard Book Number 0-309-53122-5 (PDF)

Additional copies of this report are available from the National Academies Press, 500 Fifth Street, N.W., Lockbox 285, Washington, DC 20055; (800) 624-6242 or (202) 334-3313 (in the Washington metropolitan area); Internet, http://www.nap.edu

# THE NATIONAL ACADEMIES
*Advisers to the Nation on Science, Engineering, and Medicine*

The **National Academy of Sciences** is a private, nonprofit, self-perpetuating society of distinguished scholars engaged in scientific and engineering research, dedicated to the furtherance of science and technology and to their use for the general welfare. Upon the authority of the charter granted to it by the Congress in 1863, the Academy has a mandate that requires it to advise the federal government on scientific and technical matters. Dr. Bruce M. Alberts is president of the National Academy of Sciences.

The **National Academy of Engineering** was established in 1964, under the charter of the National Academy of Sciences, as a parallel organization of outstanding engineers. It is autonomous in its administration and in the selection of its members, sharing with the National Academy of Sciences the responsibility for advising the federal government. The National Academy of Engineering also sponsors engineering programs aimed at meeting national needs, encourages education and research, and recognizes the superior achievements of engineers. Dr. Wm. A. Wulf is president of the National Academy of Engineering.

The **Institute of Medicine** was established in 1970 by the National Academy of Sciences to secure the services of eminent members of appropriate professions in the examination of policy matters pertaining to the health of the public. The Institute acts under the responsibility given to the National Academy of Sciences by its congressional charter to be an adviser to the federal government and, upon its own initiative, to identify issues of medical care, research, and education. Dr. Harvey V. Fineberg is president of the Institute of Medicine.

The **National Research Council** was organized by the National Academy of Sciences in 1916 to associate the broad community of science and technology with the Academy's purposes of furthering knowledge and advising the federal government. Functioning in accordance with general policies determined by the Academy, the Council has become the principal operating agency of both the National Academy of Sciences and the National Academy of Engineering in providing services to the government, the public, and the scientific and engineering communities. The Council is administered jointly by both Academies and the Institute of Medicine. Dr. Bruce M. Alberts and Dr. Wm. A. Wulf are chair and vice chair, respectively, of the National Research Council.

**www.national-academies.org**

# Preface

The National Research Council's Committee for Oversight and Assessment of U.S. Department of Energy Project Management has completed its assigned tasks. It was chartered in 2000 in response to continuing concern in the U.S. Congress over the Department of Energy's (DOE's) processes and procedures for managing projects. The chair expresses his appreciation to the committee members for their time, hard work, persistence, and commitment to the interests of DOE and the nation.

In its 3 years of operation, this committee has crisscrossed the country to gain input from DOE and contractor personnel at DOE offices and laboratories, as well as headquarters. The committee has had the cooperation of a wide range of DOE personnel, from the secretary through project directors and support staff in the field. It has also had input from a number of DOE contractors and from the DOE Energy Facilities Contractors' Group. The committee appreciates the time and effort of these people in providing their perspectives and insights on the issues facing DOE project management and on ways to improve DOE project performance.

The findings and recommendations of the committee in its three annual reports and two interim reports are derived in large part from the candid comments of more than 200 personnel from DOE and its contractors. The committee appreciates particularly the input from the many people in DOE who are committed to improvements in the organization's project management.

The work of this committee is a continuation of the efforts of the Committee to Assess the Policies and Practices of the Department of Energy to Design, Manage, and Procure Environmental Restoration, Waste Management, and Other

*vii*

Construction Projects (Phase II committee), and it has used its predecessor's findings and recommendations as benchmarks for measuring progress. The findings and recommendations in the Phase II report, *Improving Project Management in the Department of Energy*,[1] and the previous annual reports and interim letter reports of this committee continue to be valid. Those previous reports should be read in conjunction with this report to obtain a compete view of the status of DOE's project management, DOE's accomplishments, and the problems that still need to be resolved.

This report assesses progress in improving project management at DOE during the past 3 years, which is enough time to effect significant changes in project management. Corporations have done it in less. However, as discussed in this report, DOE has not fully committed to the steps that private corporations have had to take in order to achieve project management excellence. This report recognizes DOE's accomplishments in promulgating policies and procedures and the problems that remain to be resolved if these policies and procedures are to be implemented. In particular, some senior DOE executives have pointed out that they do not have the time to spend on project management. That may be, and the committee appreciates the competing demands on these executives' time. But if senior management does not have the time to devote to projects and does not delegate the authority to people who do, then projects are not the core priority at DOE.

This assessment is based on the belief that, for project management improvements to be effective over the long run, project management and project management improvement need a champion reporting directly to the deputy secretary of DOE. Findings and recommendations on this issue go back to the 1999 Phase II report cited above. The prognosis for progress hinges on the premise that unless or until the role of project management champion is identified at a level that demonstrates the will of DOE management to effect significant cultural change, the likelihood that project management improvements will be effective or permanent is slight.

In this report, the committee recognizes the efforts made by DOE project directors and others at various sites to solve project management problems and to handle changes created by resource deficiencies through training, internships, procedure development, and other steps at the field level. Unfortunately, although such actions have positive impacts locally on specific projects, they are too few and too isolated to stimulate widespread change within the DOE culture. Thus, the report notes that even though some process improvements have been accomplished, there is much more to be done to improve project management practices.

---

[1]National Research Council. 1999. Improving Project Management in the Department of Energy. Washington, D.C.: National Academy Press.

In December 2003, after the committee's work on this report was substantively complete, the deputy secretary confirmed the department's commitment to improving project management by creating the position of associate deputy secretary, reporting directly to the deputy secretary, with responsibilities for capital acquisition and project management. This action is a substantial step toward addressing the committee's recommendations in this report and in previous reports that DOE should have a department-wide champion for project management reporting to the deputy secretary. The committee trusts that this initiative by the deputy secretary will help achieve the permanent institutionalization of the improvements in project management made during the period of the committee's existence and will help ensure additional needed, ongoing improvements.

Kenneth F. Reinschmidt, *Chair*
Committee for Oversight and Assessment of
U.S. Department of Energy Project Management

# Acknowledgment of Reviewers

This report has been reviewed in draft form by individuals chosen for their diverse perspectives and technical expertise, in accordance with procedures approved by the National Research Council's Report Review Committee. The purpose of this independent review is to provide candid and critical comments that will assist the institution in making its published report as sound as possible and to ensure that the report meets institutional standards for objectivity, evidence, and responsiveness to the study charge. The review comments and draft manuscript remain confidential to protect the integrity of the deliberative process. We wish to thank the following individuals for their review of this report:

Philip R. Clark, Nuclear Corporation (retired),
Angelo Giambusso, Stone & Webster (retired),
Fletcher H. (Bud) Griffis, Polytechnic University,
Henry J. Hatch, U.S. Army Corp of Engineers (retired),
Martha Krebs, U.S. Department of Energy (retired),
Alan Schriesheim, Argonne National Laboratory (retired), and
Richard N. Zare, Stanford University.

Although the reviewers listed have provided many constructive comments and suggestions, they were not asked to endorse the conclusions or recommendations, nor did they see the final draft of the report before its release. The review of this report was overseen by Charles B. Duke (NAE), Xerox Research and Technology. Appointed by the National Research Council, he was responsible for making certain that an independent examination of this report was carried out in accordance with institutional procedures and that all review comments were carefully considered. Responsibility for the final content of this report rests entirely with the authoring committee and the institution.

# Contents

# Acronyms and Abbreviations

AEP        acquisition execution plan
ANL        Argonne National Laboratory

BCWS       budgeted cost of work scheduled

CD-0       critical decision 0, approval of mission need
CD-1       critical decision 1, approval of system requirements and alternatives
CD-2       critical decision 2, approval of project baseline
CDF        Collider Detector at Fermi Laboratory
CERCLA     Comprehensive Environmental Response, Compensation, and Liability
           Act
CII        Construction Industry Institute
COO        chief operating officer
COR        contracting officer's representative
CPI        cost performance index

DoD        U.S. Department of Defense
DOE        U.S. Department of Energy
DUS        dynamic underground stripping

EIR        external independent review
EM         Office of Environmental Management
ESAAB      Energy Systems Acquisition Advisory Board

ES&H        environmental safety and health
EVMS        earned value management system

FFRDC       Federally Funded Research and Development Center
FNAL        Fermi National Accelerator Laboratory
FYNSP       Future Years' Nuclear Security Plan

GAO         General Accounting Office
GFS&I       government-furnished services and items

ICE         independent cost estimate
ICPP        Integrated Construction Program Plan
ICR         independent cost review
INEEL       Idaho National Engineering and Environmental Laboratory
INTEC       Idaho Nuclear Technology and Engineering Center
IPR         internal project review
IPT         Integrated Project Team

LANL        Los Alamos National Laboratory
LBNL        Lawrence Berkeley National Laboratory
LLNL        Lawrence Livermore National Laboratory

M&O         management and operations

NA-54       NNSA Office of Project Management and Systems Support
NASA        National Aeronautics and Space Administration
NIF         National Ignition Facility
NNSA        National Nuclear Security Administration

OECM        Office of Engineering and Construction Management
OH          Ohio Field Office
OMB         Office of Management and Budget
OMBE        Office of Management and Budget Evaluation
OPC         other project costs
ORNL        Oak Ridge National Laboratory

PA&E        Office of Program Analysis and Evaluation
PARS        Project Assessment and Reporting System
PBC         performance-based contracting
PDRI        Project Definition Rating Index
PEP         Project Execution Plan
PMCDP       Project Management Career Development Program

PMP         *Project Management Practices*
PMSO        project management support office
PNNL        Pacific Northwest National Laboratory
PSO         Program Secretarial Office

SC          Office of Science
SLAC        Stanford Linear Accelerator Center
SNL         Sandia National Laboratories
SPI         schedule performance index
SRO         Savannah River Operations
SRS         Savannah River Site

TEC         total estimated cost
TPC         total project cost
TYCSP       Ten Year Comprehensive Site Plan

UT          University of Tennessee

VE          value engineering

# Executive Summary

The National Research Council's (NRC's) Committee for Oversight and
Assessment of U.S. Department of Energy Project Management has spent 3 years
(2000–2003) reviewing DOE project management policies and observing actual
practices at DOE headquarters; at a number of field sites, including Albuquerque
Operations, Oakland Operations, Richland Operations, and Oak Ridge Opera-
tions; at national laboratories, including Sandia National Laboratories (SNL),
Los Alamos National Laboratory (LANL), Stanford Linear Accelerator Center
(SLAC), Lawrence Livermore National Laboratory (LLNL), Lawrence Berkeley
National Laboratory (LBNL), and Oak Ridge National Laboratory (ORNL); at
National Nuclear Security Administration production sites at Oak Ridge and
Savannah River; and at environmental management sites at Richland and
Savannah River.

The principal goal of this effort has been to review and comment on DOE's
recent efforts to improve its project management, including a review of the following: (1) specific changes implemented by DOE to achieve improvement (e.g., in
organization, practices, policies, procedures, training); (2) an assessment of the
progress made in achieving improvement; and (3) an evaluation of the likelihood
that improvement will be permanent.

Regarding the first charge, to assess specific changes in organization, management practices, personnel training, and project reviews and reporting, the
committee finds that there has been progress in 3 years. At the time of the first of
the three assessment reports—the 1999 Phase II report (NRC, 1999)—there was
little documentation of DOE management's expectations regarding project management, if there were any. In 1999 the basic perception reported by DOE

*1*

personnel was that DOE management did not want to hear bad news. The view of the Committee to Assess the Policies and Practices of the Department of Energy to Design, Manage, and Procure Environmental Restoration, Waste Management, and Other Construction Projects (the Phase II committee) was that DOE management should define its expectations regarding acceptable project management and then document these expectations so that everyone in the organization would know what they were. Less important than the details of the expectations was the fact that DOE management had some expectations and would adhere to them. DOE management's intentions regarding policies and procedures in general were defined in 2000 by Order O 413.3 (DOE, 2000), but the requirements for implementing program and project management were not issued until 2003 in Manual M 413.3-1 (DOE, 2003). This delay is an indication that DOE management does not have a consistent set of expectations about project management across the agency. Even though the order had been issued and the decision made, considerable opposition existed internally and externally. The philosophy of successful organizations, that once the leadership has made a decision everyone unites to carry it out, has not been implemented at DOE with regard to project management.

Nonetheless the committee has observed a number of improvements in the policies that define the process by which DOE plans, selects, approves, acquires, manages, and executes projects. Notable among these process improvements are the following:

- Issuance of Order O 413.3 and its supporting manual;
- Formation of the Office of Engineering and Construction Management (OECM) and the Office of Program Analysis and Evaluations (PA&E) in the Office of Management and Budget Evaluation (OMBE);
- Formation of the project management support offices (PMSOs) in the National Nuclear Security Administration (NNSA), the Office of Environmental Management (EM), and the Office of Science (SC);
- Increased focus on strategic planning and budgeting, especially by NNSA, through its Future Years' Nuclear Security Plan (FYNSP), Ten Year Comprehensive Site Plans (TYCSPs), and Integrated Construction Program Plans (ICPPs);
- The top-to-bottom study and the strategic redirection of EM projects with the specified objectives of earlier completion at lower cost; and
- Development of the Project Management Career Development Program (PMCDP).

Additional notable accomplishments are identified more fully in Chapter 2.

Regarding the second item in the charge to the committee, the results from its assessment of progress are mixed. Concerning DOE doing the right projects to support its missions—raised as an essential point in the 1999 NRC report (NRC, 1999)—DOE has made substantial progress in defining mission requirements and long-term plans that forecast and justify the need for new projects and the priorities to be placed on them. The integration of preproject planning with long-term mission statements, if continued, should do much to advance DOE's project acquisition process.

DOE has also expended considerable effort in the development of the PMCDP, an effort that took 3 years to plan, but whether the program will be funded and fully implemented remains uncertain. In spite of the expense and complexity of its projects, DOE invests little in human resource development for project management compared with the efforts of other federal agencies or private corporations in this area. However, although DOE project directors could benefit from more professional education in the roles of the owner's representative, the problem is as much concerned with quantity as with quality. There are simply too few qualified DOE project directors and project management support staff for the number and complexity of DOE projects. The committee believes that DOE cannot afford to forgo adequate human resources devoted to project management.

Regarding the third item in the charge to the committee, to assess the likelihood that improvements will be permanent, the committee can offer little assurance. The progress cited above and documented throughout this final report is largely paper progress. The concern of the committee is not so much that Order O 413.3, Manual M 413.3-1, other documents, and the PMCDP will be rescinded, but rather that they will be circumvented. DOE's record of continual internal opposition to the order, understaffing of project directors and staff, and underfunding of project management training does not augur well for future success. The view of the committee is that if DOE were serious about continuous improvement of project management, it would put metrics in place to measure progress. However, there are no metrics in place. Even in obvious areas—for example, value engineering, which is required by Order O 413.3 and even by public law—there are no indicators to show any actual improvement in 3 years.

The committee sought out best practices in industry for comparison; it observed that a number of large industrial firms, having recognized the need for improved project management, were able to execute complete turnarounds, going from poor to excellent in project management practices in 3 years (NRC, 2002). The common factors that drove improvements in these companies are (1) a commitment from top management, (2) a strong, visible champion for project management and process improvement, and (3) a consistent, disciplined process with an emphasis on front-end planning. The case studies reviewed by the committee demonstrated that excellence in project management in industry is achieved only when the chief executive officer (CEO) or chief operating officer (COO) becomes convinced that it is essential to the success of the corporate mission, puts the resources and prestige of his or her position behind it, appoints a project management champion reporting directly to the CEO or COO, and becomes directly involved in approvals of project plans from the earliest stages. There is no shortcut or secret method, and the process is not glamorous. In these companies, commitment to the corporate position on project management becomes a condition of employment. The committee has not observed this consistent level of commitment throughout DOE. Although DOE has made progress in 3 years, it is far from a complete turnaround, and the battle is far from over.

Several factors have contributed to the slow pace of project management improvements. These include the desire of DOE personnel and contractors to be independent of oversight from DOE headquarters, slow implementation of the PMCDP and insufficient support for training, inadequate numbers of professional project directors (DOE project managers), and the absence of a champion for project managers and process improvement who is at a level of authority to be able to ensure adherence to policies and procedures and the availability of the necessary funding and personnel resources. The result of these impediments is inconsistent project performance. These issues require senior management attention to achieve progress in the future. The areas that the committee finds to be in need of additional attention are addressed in findings and recommendations in previous reports (NRC, 2001, 2003) (see Appendix D) and in the recommendations in Chapters 2 and 3.

The committee's concerns include the following:

- The momentum toward improved project management described above was attributable to the efforts undertaken by a number of influential persons within DOE in various critical management roles. Some of these people have now left DOE. Whether DOE can develop new leaders or whether the remaining leaders are or will become strong and visible champions of project management issues in DOE remains to be seen.[1]
- The committee has previously taken the position that DOE project management should be expanded and professionalized through the training of DOE project directors and supporting staff. Nevertheless, there has been internal opposition to project manager training and professional certification, and funding for the PMCDP, training courses, project management workshops, and other professional development activities has been continually in jeopardy. The amount at issue for project management professionalization is less than 0.001 percent of the amount that DOE spends on projects.
- The committee has taken the position that DOE project directors and project support staffs are inadequate in numbers to carry out the owner's

---

[1]Editor's note: The 2003 assessment is based on information reviewed by the committee through September 2003. In December 2003 the DOE deputy secretary appointed an associate deputy secretary with responsibilities for capital acquisition and project management, a positive step of which readers should be aware as they consider the committee's comments and recommendations regarding the need for a strong and visible champion of project management issues in DOE.

(DOE's) responsibilities for large and complex projects funded by tax-payer dollars. Nevertheless, the number of project management positions is apparently being cut back.

- DOE Order O 413.3 and the Energy Systems Acquisition Advisory Board (ESAAB) critical decision process have been implemented, and many project directors in the field who were skeptical 3 years ago now express the view that this decision process has added value to the project delivery system. Despite this change, efforts continue to exempt certain sites and projects from the critical management review process and to raise the threshold level of projects subject to review, even as high as $100 million.
- The congressionally mandated program of external independent reviews (EIRs) has been instituted, streamlined, and expanded. Many project directors and others comment that they have received useful information from these reviews, but others continue to deny their value. While improvements in the EIR process are possible and desirable, the committee strongly believes it would be a mistake to reduce the EIR program at this time.

The advances made in 3 years in DOE project management are fragile, and the legacy DOE culture is strong. Senior management attention and actions are essential if past improvements are to be made permanent and ingrained in the organization.

Undersecretary Robert Card has stated that, by the nature of DOE's business, excellence in the execution of large, complex projects must be a core competency of DOE. The committee wholeheartedly agrees with this view, but does not find that this goal has yet been achieved. The committee could not stress more strongly the need for continued active support from the senior leadership and staff of DOE to make that goal a reality.

Committee members talked with many people inside DOE who support improved project management. Indeed, if such individuals did not exist, improvement would be impossible. But many of these people feel that they lack support in headquarters, lack authority to carry out their duties, and lack the senior management advice and support needed to be a strong owner's representative. They need a champion to back them up, just as project managers in industry need champions in their organizations.

The deputy secretary is DOE's chief operating officer and chief acquisition executive. As such, the deputy secretary has the responsibility for assuring that projects are effectively planned and executed. To perform these functions, the deputy secretary himself or herself either should be the champion for project management improvement—to develop project management into a core competency of the department, to assure that the department maintains an adequate staff of qualified project directors to manage its portfolio of projects, and to assure that the disciplined execution of projects is a priority for managers at all levels—or should appoint someone to perform this role, reporting directly to the deputy secretary (see footnote on page 4). Senior DOE managers have shown visible support for policies and procedures and the ability to step in to resolve deadlocks, but these actions are the equivalent of short-term fire fighting, which may be insufficient to sustain continued process improvement. The escalation of organizational deadlocks and internal disputes up to senior management for resolution is an indication of how the system is not working. The new policies and procedures demonstrate substantial progress in DOE, but the committee is not confident that these changes will be permanent without a strong champion to implement and solidify these procedures across the complex (see footnote on page 4).

## REFERENCES

DOE (U.S. Department of Energy). 2000. Program and Project Management for the Acquisition of Capital Assets (Order O 413.3). Washington, D.C.: Department of Energy.

DOE. 2003. Project Management for the Acquisition of Capital Assets (Manual M 413.3-1). Washington, D.C.: Department of Energy.

NRC (National Research Council). 1999. Improving Project Management in the Department of Energy. Washington, D.C.: National Academy Press.

NRC. 2001. Progress in Improving Project Management at the Department of Energy, 2001 Assessment. Washington, D.C.: National Academy Press.

NRC. 2002. Proceedings of Government/Industry Forum: The Owner's Role in Project Management and Preproject Planning. Washington, D.C.: National Academy Press.

NRC. 2003. Progress in Improving Project Management at the Department of Energy, 2002 Assessment. Washington, D.C.: The National Academies Press.

# 1

# Introduction

## BACKGROUND

The Department of Energy's (DOE's) diverse missions are supported by scores of projects, resulting in annual expenditures of billions of dollars. Consequently, the U.S. Congress has an ongoing concern about project management in DOE and the need to assure American taxpayers that the nation's resources are effectively and efficiently managed. In response to a directive from the Committee of Conference on Energy and Water Development of the 106th Congress (U.S. Congress, 1999), DOE requested that the National Research Council (NRC) appoint a committee to review and assess the progress made by the department in improving its project management practices. The principal goal of this effort has been to review and comment on DOE's recent efforts to improve its project management, including a review of the following: specific changes implemented by DOE in order to achieve improvement (e.g., in organization, practices, training); an assessment of the progress made in achieving improvement; and an evaluation of the likelihood that improvement will be permanent. (See Appendix A for the statement of task.)

This oversight and assessment is the third phase of evaluative activities that began in 1997. The first phase was an assessment of the need for independent project reviews (Phase I) (NRC, 1998), which was followed in 1998 by a comprehensive assessment of project management practices (Phase II) (NRC, 1999). The Phase III assessment was planned as a 3-year effort beginning in July 2000 and includes three annual reports, as described below.

The NRC appointed a committee under the auspices of the Board on Infrastructure and the Constructed Environment (BICE) to undertake the review and assessment of DOE project management. The committee is composed of 10 professionals with diverse experience in academic, government, and industrial settings and extensive knowledge of project management and process improvement. Three members of the committee also participated in the Phase II review and assessment, and one member participated in both Phase I and Phase II efforts. (See Appendix B for biographies of the committee members.)

This document is the committee's third and final annual report. It includes the committee's assessment of progress in improving project management at DOE as of September 2003 and provides additional discussion of what the committee determined to be key factors affecting DOE project management.[1] The body of this report addresses some of the issues raised in the Phase II report (NRC, 1999), and provides a continuation and summation of the assessments published in the committee's 2001 annual report (NRC, 2001b); 2002 annual report (NRC, 2003); interim letter reports (NRC, 2001a, 2002a); and proceedings of the 2001 forum on the owner's role in project management and preproject planning (NRC, 2002b). Not all of the findings and recommendations in the previous reports are specifically cited here, although the committee continues to endorse them.

This assessment has focused on the program activities in the Office of Environmental Management (EM), the National Nuclear Security Administration (NNSA), and the Office of Science (SC) because they are responsible for the majority of DOE projects. The committee notes that other DOE program offices, such as the Office of Nuclear Energy Sciences and Technology and the Office of Civilian Radioactive Waste, have significant project responsibilities. The committee's findings and recommendations for disciplined and effective project management address the need for a consistent process and level of performance for all projects undertaken by all program offices in DOE.

## REVIEW ACTIVITIES

Beginning in September 2000, the committee met 14 times to review and assess the data on projects and project management procedures presented by the DOE project managers and representatives of the Office of Management and Budget Evaluation (OMBE), the Office of Engineering and Construction Management (OECM), and the project management support offices (PMSOs) in EM, NNSA, and SC. The committee also met with DOE personnel and DOE contrac-

---

[1]In December 2003, the DOE deputy secretary appointed an associate deputy secretary with responsibilities for capital acquisition and project management. This appointment has been noted where it has an effect on the committee's assessment.

tor personnel in Albuquerque, New Mexico; Berkeley, Oakland, Livermore, and Stanford, California; Oak Ridge, Tennessee; Richland, Washington; and Aiken, South Carolina. Committee representatives also attended project management workshops and awards programs sponsored by OECM, the PMSOs, and nongovernmental organizations in 2000 and 2001. Committee representatives have met with the secretary of energy, the deputy secretary, the undersecretaries and assistant secretaries, the director of OMBE, and other DOE senior managers responsible for managing programs, establishing policies, and implementing project management reforms.

The committee received input from more than 200 personnel from DOE, contractors, and other government agencies (e.g., the General Accounting Office [GAO] and congressional staff), often on multiple occasions, in order to assess changes in their views and attitudes concerning DOE project management. From personal meetings, briefings, and conference participation, the committee received a wide range of views on how to improve project management in DOE.

The committee's observations, findings, and recommendations are derived in part from comments and suggestions made by DOE personnel and DOE contractors. The committee's fact-finding efforts from February 2003 through June 2003 are listed in Appendix C. Previous fact-finding efforts are listed in the earlier reports.

## ORGANIZATION OF THE REPORT

This report includes an evaluation of the implementation of recommendations in the Phase II report (NRC, 1999) and the specific changes in organization, management practices, personnel training, and project reviews and reporting since publication of the Phase II report. It provides the committee's assessment of the progress made in improving project management and the likelihood that improvements will be permanent. The report is organized in three chapters and seven appendixes.

Following the Executive Summary and this chapter's presentation of background information on the initiation and conduct of the study, Chapter 2, "Progress and Opportunities," describes DOE's accomplishments in improving project management and the issues that the committee believes need further attention and improvement. The chapter addresses organizational changes, policies, and procedures that have been issued, human capital, cultural change, project planning, risk management, project controls, performance metrics, project reviews, and acquisition and contracting. Guidance for future improvements is provided in the committee recommendations.

Chapter 3, "Prognosis for Progress," addresses the likelihood that project management improvements are sustainable and will continue to improve. The findings and recommendations address the issues that are most critical to the future of DOE project management.

The report's seven appendixes provide additional background information to support the committee's assessment: the statement of task; biographies of committee members; a list of the fact-finding sessions, briefings, and documents reviewed since the 2002 annual report; a compilation of previous findings and recommendations from Phase II and Phase III reports; a memorandum of April 2000 from the Undersecretary regarding the delegation of acquisition executive authority; a summary of key findings in recent external independent reviews; and correspondence between the DOE Office of Science and the National Academies regarding the 2002 assessment report.

## REFERENCES

NRC (National Research Council). 1998. Assessing the Need for Independent Project Reviews in the Department of Energy. Washington, D.C.: National Academy Press.

NRC. 1999. Improving Project Management in the Department of Energy. Washington, D.C.: National Academy Press.

NRC. 2001a. Improved Project Management in the Department of Energy. Letter report, January. Washington, D.C.: National Academy Press.

NRC. 2001b. Progress in Improving Project Management at the Department of Energy, 2001 Assessment. Washington, D.C.: National Academy Press.

NRC. 2002a. Progress in Improving Project Management at the Department of Energy, 2002 Interim Assessment. Letter report, May. Washington, D.C.: National Academy Press.

NRC. 2002b. Proceedings of Government/Industry Forum: The Owner's Role in Project Management and Preproject Planning. Washington, D.C.: National Academy Press.

NRC. 2003. Progress in Improving Project Management at the Department of Energy, 2002 Assessment. Washington, D.C.: The National Academies Press.

U.S. Congress. 1999. House of Representatives, Energy and Water Appropriations Bill, 2000. HR 106-253. Washington, D.C.: Government Printing Office.

# 2

# Progress and Opportunities

## INTRODUCTION

The 1999 NRC report entitled *Improving Project Management in the Department of Energy* provided guidelines for "lifting DOE's project management to a level commensurate with other agencies and private industry" (NRC, 1999, p. 3). This chapter provides the committee's observations of DOE's accomplishments in implementing the recommendations in the 1999 report and in subsequent assessment reports from 2001 through 2003 (NRC, 1999, 2001a, 2001b, 2002a, 2002b, 2003). The committee has observed significant progress over the past 3 years in the improvement of project management at DOE, but action on most of the committee's past recommendations is still incomplete, and the results department-wide are inconsistent. Progress has been demonstrated in the areas of administrative organization, departmental policies and manuals, management directives, a career development program, and changes in attitude of many DOE personnel. The committee also notes that, despite these actions, the need for additional improvement continues, and that significant commitment and additional effort will be needed to ensure consistent implementation of the improved policies and procedures.

DOE projects are often very large and technically challenging. Because these projects are so important and so costly, the committee believes that DOE's goal should be to improve its project management to a level at least commensurate with that in other agencies and private industry and that DOE should strive to achieve a level of excellence commensurate with its accomplishments in science and defense technology. In its previous annual reports (NRC, 2001b, 2003), the committee identified specific aspects of project management at DOE that need improvement in order to bring the department's project management procedures and project performance to an acceptable level of competence. The three previous NRC reports (NRC, 1999, 2001b, 2003) include 76 findings and 94 recommendations that the committee believes are still applicable. (See Appendix D for a compilation of findings and recommendations.) The previous findings and recommendations as well as those in this report address 10 recurring objectives for the changes needed to improve project management at DOE. The recurring objectives include the following:

- Develop policies and procedures to define the DOE method of managing projects;
- Create a project management culture across the agency that supports the consistent implementation of policies and procedures;
- Provide leadership that ensures disciplined planning and execution of projects as well as support for continuous process improvement;
- Provide a project management champion at the highest level of the department to ensure that a focus on the importance of project management is established and maintained;
- Develop competence in fulfilling the owner's role in strategic planning, front-end project planning, risk management, and project execution;
- Apply rigorous project reporting and controls that include earned value systems; link day-to-day management data to periodic reporting and forecast time and cost to complete; and maintain historical data with which to benchmark project performance;
- Document processes and performance to support benchmarking and trend analysis;
- Invest in human capital by providing training and career development to ensure an adequate supply of qualified, skilled project directors and support staff;
- Continue, refine, and document a program of external and internal project reviews; and
- Employ innovative approaches to capital acquisition and the use of performance-based contracting.

This chapter provides the committee's assessment of DOE's progress in achieving these objectives and actions needed to continue progress toward an appropriate level of excellence in project management. Although the committee has provided suggested approaches, it believes that DOE senior managers have the responsibility to identify and apply metrics to define the appropriate level of excellence and to drive continuous process improvement.

# ORGANIZATION

Organizational changes in DOE in the past 3 years have established a number of administrative offices specifically to manage projects more effectively and to improve project outcomes. These include the Office of Engineering and Construction Management (OECM) and the Office of Program Analysis and Evaluation (PA&E) in the Office of Management and Budget Evaluation (OMBE) and the project management support offices (PMSOs) within the major programs. The committee notes also that the leadership and involvement of senior DOE managers are key factors in the success of project management improvement efforts.

## Office of Engineering and Construction Management

The OECM was established to implement project management reforms initiated in June 1999 (DOE, 1999). DOE's *Program and Project Management Policy for the Planning, Programming, and Acquisition of Capital Assets* (P 413.1) established OECM in the Office of the Chief Financial Officer, which was later reorganized as the Office of Management and Budget Evaluation (DOE, 2000a).

The OECM mission is to provide the department with consistent, reliable project management processes, to drive improvements in project management systems, to integrate sound acquisition and business practices, to support the professional development of project managers, and to facilitate senior managers' oversight of the department's projects (DOE, 2003d). In the past 3 years, OECM developed *Program and Project Management for the Acquisition of Capital Assets* (Order O 413.3) (DOE, 2000b); *Project Management for the Acquisition of Capital Assets* (Manual M 413.3-1) (DOE, 2003a); *Project Management Practices* (DOE, 2003f); the Project Management Career Development Program (PMCDP) (DOE, 2003b); and the Project Assessment and Reporting System (PARS) (DOE, 2003d).

OECM manages and coordinates external independent reviews (EIRs) of projects, reviews acquisition plans, and serves as the secretariat for the Energy Systems Acquisition Advisory Boards (ESAABs). The committee believes that OECM is providing a vital quality-assurance function by providing DOE senior management with the information and advice essential to determine if and when a project is ready to proceed to the next step, and if projects are appropriately planned and executed. The committee also believes that OECM is at a level in the organization that it can be effective in improving project management if it is adequately staffed, senior management is sufficiently committed and involved, and senior management uses OECM and other resources to positively influence project management discipline across DOE.

## Office of Program Analysis and Evaluation

The Office of Program Analysis and Evaluation, like OECM, reports to the OMBE director. PA&E was established in 2001 to provide independent analytical advice to DOE acquisition executives regarding the planning, execution, evaluation, and measurement of the effectiveness of DOE missions and programs by developing the department-wide strategic management system (DOE, 2003c). PA&E has helped to improve project management by evaluating proposed projects in order to assess their alignment with the department's mission and strategic plan. The committee believes that PA&E can provide a vital function by facilitating senior management's determination that the department is funding the right projects with the appropriate priority and level of funding.

## Project Management Support Offices

The majority of DOE projects are accomplished by three program organizations—the Office of Environmental Management (EM), the National Nuclear Security Administration (NNSA), and the Office of Science (SC). Each of these organizations has established a project management support office that provides guidance for project management procedures, supports internal project reviews (IPRs) and internal reporting, and coordinates project manager training and certification, ESAAB critical decision processes, and senior management reporting and oversight with OECM. The PMSOs, though generally understaffed in the committee's judgment, have played an important role in advocating and implementing project management improvement within their respective program offices.

## Senior Management

The committee believes that PMSOs, PA&E, and OECM can be effective in improving the management and execution of DOE projects only if they are used and supported by senior managers. To ensure that these offices are effective, the deputy secretary, undersecretaries, and assistant secretaries should do the following:

- Actively use PA&E to provide objective analysis and advice on mission need and project cost-benefit justification at the early stages of incipient projects;
- Actively use OECM to develop and promulgate requirements for management decisions, monitor the progress of projects, and provide objective analysis and advice concerning project management through all project reviews and critical decision points; and
- Strengthen these functions by providing adequate resources, staffing, and training and by empowering OECM, PA&E, the PMSOs, and DOE project directors and project support staffs.

## Project Management Leadership

The committee has been impressed by examples of management leadership at the highest levels of DOE in implementing improvements to project management practices. The committee believes that senior management at DOE, especially the deputy secretary, needs to be highly visible in promoting project management excellence to ensure that it will succeed and that improvements will be permanent.

The committee believes that sustainable competence in project management at DOE can only come through the efforts of senior management with the authority to ensure agency-wide compliance with the project management procedures. In the 2001 assessment, the committee emphasized the need for intensive senior management involvement. Key actions by senior managers—for example, involvement in reviews, ensuring that adequate resources are provided, and ensuring process discipline—are critical to long-term improvement (NRC, 2001b). The committee has observed instances of effective leadership, but the leadership has been neither consistent enough nor rigorous enough to ensure the continued improvement of project management. For example, the committee notes that the direct intervention of the undersecretary and deputy secretary was necessary to achieve consensus on the project management manual so that it could be issued. However, it took almost 3 years for DOE to resolve the internal differences and issue this manual, indicating that DOE management does not have a clear strategy on how to manage projects. This example amounts to short-term firefighting, and a more sustained, visible commitment by senior managers is needed in order to continue improving project management and to make these improvements a permanent part of DOE.

Industrial organizations that have created excellent internal project control organizations have typically assigned the responsibility for project management programs to managers at the level of vice president or senior vice president. These senior executives, and even corporate CEOs, find it essential for success to maintain direct cognizance over projects—even those costing as little as $5 million. In DOE, the deputy secretary, as the department's COO and senior acquisition executive, is responsible for effective project management. The deputy secretary should be DOE's champion of project management excellence. As noted in previous NRC reports, the committee believes that DOE needs a visible, active, senior-level manager to promote and defend efforts to improve project management capabilities and their consistent application throughout the department (NRC, 1999).

The committee also recognizes that DOE program organizations are given independent authority and accountability for managing projects. The assistant secretaries for SC and EM have been delegated the authority by the undersecretary to act as the acquisition executives for projects under $400 million in their respective programs. The NNSA administrator, as DOE undersecretary, also has this same authority.  A memorandum from Undersecretary Robert Card to the assistant secretaries notes that they are also accountable for the quality of their respective programs' project management systems (see Appendix E).  The committee believes that just as the deputy secretary should be a champion for project management throughout the department, the program heads should be strong, visible champions for project management in their programs.

Although it is essential that each program office have strong project management capabilities and support, the committee continues to recommend that it is necessary that there be a single entity responsible for project management policies, procedures, personnel career development and training, and project tracking and reporting in order to achieve consistent project management excellence across the agency.

A number of DOE project directors expressed the belief that DOE upper management would not support them if they rejected contractors' submittals or decisions, and they were thus reluctant to challenge contractors.  Orders, manuals, and guidelines are necessary but not sufficient—project directors need to know that they have someone to whom they can turn for professional guidance and support when making difficult decisions.  To have the necessary impact to affect the DOE culture, this function should report directly to the deputy secretary.[1]

## POLICIES AND PROCEDURES

Considerable effort has been devoted to developing project management policies and procedures over the past 3 years.  An order entitled *Program and Project Management for the Acquisition of Capital Assets* (Order O 413.3); the manual *Project Management for the Acquisition of Capital Assets* (M 413.3-1); a guide entitled *Project Management Practices* (PMP); and the notice *Value Engineering* (N 413.2) have been developed (DOE, 2000b, 2002, 2003a, 2003d).

### Program and Project Management Policies for the Planning, Programming, and Acquisition of Capital Assets

DOE Policy P 413.1 (DOE, 2000a) set the stage for revitalizing project management within DOE.  It charged OECM with the responsibility of preparing project management procedures for the department.  DOE Order O 413.3 (DOE, 2000b), issued in October 2000, defines the principles of DOE project manage-

---

[1]Editor's note:  The 2003 assessment is based on information reviewed by the committee through September 2003.  In December 2003 the DOE deputy secretary appointed an associate deputy secretary with responsibilities for capital acquisition and project management, a positive step of which readers should be aware as they consider the committee's comments and recommendations regarding the need for a strong and visible champion of project management issues in DOE.

ment and project oversight. It includes requirements for Integrated Project Teams (IPTs) and defines a critical decision process to assure the involvement of senior management as responsible decision makers from the inception to the completion of projects. In addition, the oversight process is enhanced by the ESAABs, which advise the acquisition executives at critical decision points.

Following the issuance of Order O 413.3, OECM conducted workshops to obtain feedback and build support among DOE project managers. The order was reinforced by memoranda issued by Francis Blake, deputy secretary of DOE, in September and November 2001, and subsequently by Bruce Carnes, director of OMBE, in February 2002. The committee applauds both the order and the efforts to implement it throughout the department. The committee understands that the order may need to be updated from time to time, but believes that the basic principles, procedures, and applicability should be kept intact.

## Project Management for the Acquisition of Capital Assets Manual

Considerable effort went into preparing a manual for the implementation of Order O 413.3. The first draft was issued in October 2000, and revised drafts were issued in February, June, and August 2002. Each iteration responded to critical review feedback from DOE project managers (project directors), contractors, this committee, and others. The final document's format and organization present the roles, responsibilities, limits of authority, and required project controls and review procedures with respect to project management for the acquisition of capital assets. Efforts to issue the manual were threatened by internal and external resistance, but contentious issues were finally resolved through strong leadership and direct intervention by the undersecretary and the persistent efforts of OECM and the PMSOs. Manual M 413.3-1 was finally published in March 2003 (DOE, 2003a). Further, the deputy secretary issued a memorandum directing each Program Secretarial Office (PSO) to hold implementation sessions at field offices to demonstrate the commitment of line management to the project management system outlined by the manual. The committee was pleased to see this direct involvement of top DOE leadership in efforts to improve project management. This episode shows that attention by senior management will continue to be necessary to achieve the cultural change necessary to institutionalize project management best practices throughout the department.

The overall structure and integration of M 413.3-1 as published have been considerably improved in comparison with previous versions. The document does an admirable job of describing the key issues that need to be addressed for project initiation—that is, approval of mission need (critical decision 0 [CD-0]) and approval of system requirements and alternatives (CD-1), including determining the estimated cost range. However, the document provides little direction or guidance with regard to the critical tasks that must be performed in the period between CD-1 and the approval of the project baseline (CD-2). This important front-end planning phase of a project is mentioned only on a few pages (that is, on pages 2-5, 2-6, 6-1, 6-2; at CD-2 review on pages 9-6 and 9-7; and again briefly in Chapter 10). In contrast, the CD-0 and CD-1 deliverables merit an entire chapter each (Chapters 4 and 5, respectively).

Since the activities between CD-1 and CD-2 typically are used to define the detailed scope of the project (as opposed to the conceptual scope) and encompass critical functions such as site evaluation, process flow design, design parameters, coordination issues, safety, instrument and electrical diagrams, equipment scope, execution plans, and so on, the committee believes this to be a significant omission from the manual. As noted in the 2001 forum on preproject planning, these functions are critical to the front-end planning process in successful project management organizations, and they can significantly affect project performance (NRC, 2002b).

Although contractors will typically perform these activities on DOE projects with oversight by a federal project director, it is imperative that the department outline specific guidance on the detailed deliverables needed so as to ensure that the activities are done properly. In addition, federal staff needs to have sufficient management and technical expertise to assess the quality of the contractor's project management products.

The committee believes that the lack of explicit guidance for project management oversight between CD-1 and CD-2 exemplifies the absence of a clear understanding of the owner's role in the project planning process. The key findings in the external independent reviews for the Center for Nanophase Materials Sciences, the Oak Ridge Research Support Center, and the Ashtabula Closure Project illustrate the problems that are still prevalent on many projects (see Appendix F). Both the manual and *Project Management Practices* need to be expanded to discuss the procedures and deliverables required at this phase, and federal project directors need to be involved in this effort to incorporate best industry practices.

## Project Management Practices

*Project Management Practices* (PMP) consists of guidelines issued as references for project directors. The PMP elaborates on the information contained in Order O 413.3 and Manual M 413.3-1 by providing supplementary instructions that, although not required, are recommended in order to improve DOE's ability to manage projects (DOE, 2003d). The PMP is distributed on CD-ROM and on the World Wide Web in order to permit frequent updating. The committee endorses issuance of the practices as a useful tool for present and future DOE project directors.

## Value Engineering

Deficiencies in DOE's value engineering (VE) efforts were noted in the Phase II report (NRC, 1999) and the 2001 letter report (NRC, 2001a), and VE procedures were cited as being in need of additional documentation in the 2001 assessment (NRC, 2001b). DOE has made a considerable effort to make known to project directors the federal requirements to perform value engineering, primarily through a Web page listing of federal statutes, Office of Management and Budget (OMB) circulars, and DOE policies and procedures; this Web page is linked to the OECM Web site. DOE issued VE program implementation policies and procedures, including a VE policy notice (Notice N 413.2), in December 2002 (DOE, 2002). The notice was followed by an implementing memorandum in January 2003 and a VE acquisition letter (No. 2003-04) in August 2003 (DOE, 2003e) that provides direction and guidance on the application of VE for management and operations (M&O) contracts and other contracts for the performance of work at DOE sites and facilities. VE requirements are also addressed in Chapters 2 and 5 of Manual M 413.3-1. Training for VE is included in the Project Management Career Development Program (PMCDP) as an elective. Work is under way to include guidance on VE in the PMP and to develop a Web-based tutorial.

Policies are now in place, but the committee cannot determine if they are being implemented, because performance measures are not yet available for assessing actual progress. The committee is aware of $16 million in VE savings in NNSA for 2002 but has no information on savings in other programs. It appears that DOE continues to lag behind the Department of Defense (DoD) and other federal agencies in implementing established, government-wide VE requirements. DOE appears to be taking action to implement a VE program, but additional attention and resources are needed to make VE an integral part of project management.

DOE's VE program exemplifies the committee's concerns with DOE: VE is not new; it is not mysterious, exotic, or difficult; it is routinely practiced by other federal agencies; it is mandated by law; and it has been required by O 413.3 since 2000. Yet DOE has not effectively implemented VE. The apparent lack of commitment to implementing VE reduces the committee's confidence in DOE's long-term commitments to implementing policies and procedures and improving project management.

## Owner's Role in Project Management

The committee has emphasized the need for the federal employees involved in project management functions to assume the role of the owner's representative (NRC, 1999, 2001b, 2003). In its 2002 assessment the committee made the following recommendation:

In order for DOE to be an effective owner of capital acquisition projects it should:

- Consider capital projects critical to organizational success.
- Require senior management involvement in project decision making, usually at the $5 million and higher level.
- Have a detailed and well-recognized internal front-end planning process.
- Capture metrics on planning effort and project performance.
- Require owner involvement and leadership in front-end planning.
- Ensure that projects support DOE's mission and are consistent with DOE's strategic plan. (NRC, 2003, p. 31)

As noted in other sections of this report, DOE's performance is inconsistent in activities that are critical to the role of an effective owner.

The change of the title "federal project manager" to "federal project director" may help differentiate the roles and responsibilities of contractor personnel and help reduce confusion and focus federal employees on their appropriate roles (DOE, 2003a). However, Manual M 413.3-1 contains an inadequate definition of the roles and responsibilities of DOE personnel and contractors. The definition of roles and responsibilities should follow the requirements of O 413.3. It should cover the DOE chain of responsibilities from the acquisition executive to the project director, and the contractor's chain of responsibilities from executives to managers in the field. The definitions should define responsibilities to develop, review, comment, approve, and execute at each step of the DOE capital acquisition process. These definitions should do more to address the authority that DOE project directors have in carrying out these responsibilities. In addition, each of the PSOs needs to develop a detailed "roles and responsibilities" document to reflect the different procedures used by their separate organizations. The committee notes that the process of developing this detailed information for the manual will help DOE identify opportunities to streamline the process.

DOE project directors in the field are asking for better definition of their roles, responsibilities, and authority. In particular, they want to know if anyone in DOE will support them if they make a decision that a contractor does not like.

## Applicability of Policies and Procedures

Order O 413.3 applies to all budget line-item projects over $5 million total project cost (TPC). The committee believes that, with provisions for tailoring requirements to the complexity of a project, this is an appropriate level. Because Congress wants DOE to control all of its projects, the general applicability of project management policies and procedures should remain at this level. Waivers of designated sites or projects from the requirements of O 413.3 (e.g., the Advanced Reactor Hydrogen Co-Generation Project [U.S. Senate, 2003]) are viewed by the committee as dangerous precedents that, if repeated, will undermine DOE's ability to establish a department-wide culture that can manage projects well.

Some personnel in DOE and M&O contractors in the field expressed their concerns to the committee that the O 413.3 and M 413.3-1 are overly prescriptive and that the threshold value of projects that must comply with the order is set too low. These persons expressed the opinion that compliance with the order would result in project delays caused by the involvement of senior management at the critical decision points. The committee considered these positions thoroughly and believes they lack both basis and merit. The committee saw no actual evidence of projects that were delayed during the CD-0 through CD-2 reviews. On the contrary, the committee observed many projects that were commenced under O 413.3 and that proceeded expeditiously and on schedule.

The committee noted in its previous reports the absence of adequate project justification and front-end planning, manifested in the lack of documentation such as acquisition plans, risk management plans, and project execution plans. If inadequate front-end planning documents have been returned for more justification and correction in order to support management decisions, the committee does not regard such action as a delay but as a desirable outcome.

The committee reiterates its view that doing adequate planning up front is an essential activity and should be included in the project schedule. Projects may be spending more time and effort on planning than they did before O 413.3 because, as the committee has noted in previous assessments, thorough project planning was not being performed. If a project schedule allots insufficient time for adequate planning, it is a fault in the schedule, not evidence of delay. On the basis of best practices in industry and in other government agencies, the committee expects that the time taken for adequate planning to support management decisions at CD-0 and CD-1 can be made up at later CD points, which will go more smoothly if CD-0 and CD-1 are done correctly.

If projects are actually delayed because the project justification and planning documents were inadequate to support DOE management decisions, the solution is not to eliminate management decision points but to improve the planning and the documentation. Review of project justification is an essential management quality-control point. The summary of EIR findings (see Appendix F) indicates that quality-control planning documents continue to be necessary. The committee believes that the solution is to improve quality, not to eliminate the quality-control function. If delays are indeed due to a lack of resources for OMBE and the ESAABs to perform reviews expeditiously, then more resources should be added to those activities. It is an appropriate exercise of the senior management function to delay some projects if they are considered to be marginal or perhaps unnecessary or of low priority, and to send them back for further analysis, justification, and documentation of mission need. In the past, GAO noted DOE's history of executing projects that turned out to have little value (GAO, 1998). DOE has made progress in the area of documenting and justifying projects and should not backslide. The objective of a good project management process is to execute the right projects well, not to build the wrong projects faster.

The critical management decision process is an integral part of project development in any successful organization; it should be part of DOE's. Management reviews are not delays—they are critical steps in a project's development path. The committee notes that the requirements for project justification are, in the end, whatever DOE senior management needs in order to make informed decisions. If DOE senior management wishes to avoid project mistakes, it will require adequate project justification, especially at CD-0 and CD-1. The real issue here is not procedures in O 413.3, but whether DOE senior management is able to control DOE projects and to assure that they are aligned with DOE's mission needs.

The committee has observed project documents with a full page of management sign-offs and recognizes that there are opportunities to streamline the review process by reducing the number of internal sign-offs required for ESAAB reviews. The committee recommends that the PSOs reengineer the ESAAB process to eliminate requirements for sign-offs that are not absolutely essential and do not add value to the project.

## HUMAN CAPITAL AND PROFESSIONAL DEVELOPMENT

### Project Management Career Development Program

The committee endorses the concept and contents of the Project Management Career Development Program, which is now Attachment 4 of Acquisition Career Development Program (Order O 361.1) (DOE, 2003b). The committee commends the deputy secretary for his support of the program and recommends that the department maintain its commitment to implementing the plan, which calls for education, certification, and demonstration of proven project management abilities on successively more challenging projects. However, the committee finds that after 3 years of development, the PMCDP has not been widely implemented. The committee believes that the reasons for this slow rate of progress are the lack of emphasis and low priority assigned to this program and that this slowness may reflect an underlying culture that resists change within the department. Holding back funding for personnel and training and an unwillingness to expend resources (monetary and otherwise) to support the project management team is completely inappropriate for an organization that has an ongoing capital acquisition program in excess of $40 billion.

### Project Management Professional Certification

Although professional project manager certification does not guarantee success in itself, the committee continues to emphasize it as an indicator of the importance that DOE places (or does not place) on professional education, development, and credentials for project directors. Firms in the private sector and other federal agencies such as the National Aeronautics and Space Administration

(NASA) and DoD routinely certify project managers; indeed, some organizations in DOE have done it, and there is no reason why DOE should not do it throughout the department. At least one DOE organization, NNSA's Office of Project Management and Systems Support (NA-54), has achieved 100 percent PMP certification, and others have established this as a goal, sending the message that at least part of DOE management considers project management to be a profession with professional standards. The efforts expended to become certified will enhance a project director's capabilities and provide a goal for aspiring project directors to achieve. PMP certification should be considered one means of developing project management competency within DOE. The deputy secretary's commitment to full certification of DOE project directors in all PSOs is a significant step forward.

## Training

Project management training at DOE continues to progress very slowly, with relatively few courses being offered to personnel in the field. Uncertain funding commitments have contributed to this slow start and have jeopardized full implementation of the PMCDP. The committee believes that central funding for course development, tuition, and travel expenses is essential to fostering a viable training program and that elimination of centralized funding would be tantamount to eliminating the program. The committee understands that the central working capital fund for project director training and career development has been established through FY 2005. The committee commends DOE management for recognizing the value of this essential mission-critical function.

The committee is encouraged by the fact that some field units, such as Los Alamos National Laboratory (LANL), have taken the initiative to provide needed career development training and just-in-time training for project teams, and to establish internships to develop the project directors who will be needed on current as well as future projects. These frontline personnel appreciate the immediate needs for training and professional project management certification, despite the long period of time required to develop a DOE-wide program.

The committee sees particularly urgent needs for department-wide training in front-end planning, risk analysis, and project controls—all aspects of project management that greatly need improved consistency. To make the most efficient use of training resources and to maximize staff access to training opportunities, DOE should utilize alternatives to traditional classroom delivery of instructional materials.

## Career Development

The committee is concerned that the PMCDP focuses only on current project directors and does not identify career paths for personnel aspiring to become project directors, nor does it address training and development for project management support staff. The committee believes that a comprehensive career development program is key to improving the project management culture in DOE. The best-performing organizations consistently train and promote from within (Collins, 1994). This point has been taken to heart by other federal agencies with large project responsibilities and budget—for example, NASA, which has an Academy for Program and Project Leadership, and DoD, which established the Defense Acquisition University to educate federal project management personnel.

## Project Management Workshops

The committee has recognized the value of project management workshops sponsored by OECM and the PMSOs in past reports (NRC, 2001a, 2001b, 2002b, 2003). Unfortunately, funding decisions resulted in cancellation of these workshops. The committee believes that these workshops serve a valuable function by enhancing the visibility of project management capabilities and providing opportunities for sharing lessons learned across projects and across programs. DOE project management culture is significantly enhanced by assembling a large group of personnel across the department to witness senior management's commitment to project management by the presentation of awards for outstanding project skills and project performance. Therefore, the committee reiterates the recommendation that these workshops should be continued and expanded, not terminated.

## Authority and Responsibility

The committee heard from project directors in the field that they lacked sufficient authority to discharge the responsibilities assigned by current project management policies. The project directors believe that they need to be designated as DOE contracting officer's representatives (CORs) in order to have the authority to accept or reject a contractor's project management documents and project performance reports. The committee supports appointing project directors as CORs and believes that this authority will strengthen their abilities to function effectively as owner's representatives. The committee does not advocate the development of detailed job descriptions defining the authority of project directors, although it is agreed that this authority is currently ambiguous. The committee believes that the single most effective way to strengthen the authority of the DOE project directors would be to provide them with a single mentor, supporter, and champion for project management across the complex.

## Management Resources

The committee reiterates its finding and deep concern that there are too few project directors and support staff personnel to properly discharge DOE's ownership responsibilities, considering the size and complexity of the programs. The committee believes that it is imperative for DOE to develop an adequate human resource base. The inclusion of personnel assigned to project management activities within the program direction account hinders the rational application of federal personnel resources to varying workloads. However, the most serious consequence of funding project management in the program direction account is that, as a result, DOE project management has been chronically and seriously underfunded and understaffed. Therefore, the committee recommends that project directors and project management functions be funded from project appropriations. The use of project funds for project management personnel (as is done within the Department of Defense) could be very effective in improving project management in DOE.

### Cross-Utilization

The 2002 assessment report (NRC, 2003) recommended that project managers (project directors) and project management staff be considered a department-wide resource rather than being confined to program offices. The committee has detected no significant acceptance of this principle, and it again recommends that senior managers support this approach to using scarce resources more effectively across the complex.

## RECOGNIZED VALUE OF CHANGE

The 1999 NRC report noted that the DOE culture was not conducive to effective project management and execution and that fundamental changes in the organization's culture were needed to improve performance (NRC, 1999). The following strategies were listed as possible means for creating a project management culture that would be more conducive to project management excellence than the situation extant in 1999:

The cultural change "levers" available to the secretary are the recommendations throughout this report, including the following:

- Create a culture of excellence in project management and execution.
- Establish the goal of becoming a leader in project management skills, methodology, technology, systems, and performance.
- Promulgate clear directions on project management policy, stressing that completion of projects to scope, on time, and on budget is of the highest priority.
- Provide clear definitions of responsibility, authority, and accountability for all personnel involved in projects. Prohibit interference from outside the chain of responsibility. Clarify DOE field office and contractor roles, responsibilities, authorities, and relationships.

- Enhance preconstruction planning, so that scope definition, baselines, budgets, contingencies, and schedules are realistic, and everyone involved understands what will be done, and when. After budgets are fixed, design and construct the project to meet the budget.
- Engage user managers early and require that users be committed to project scope, requirements, budget, and schedule.
- Ensure that user/client decisions are made in a timely manner to avoid project delays.
- Provide objective, standard methods for assessing project risks and uncertainties, and assign realistic budgets, schedules, and contingencies.
- Give the assigned project manager authority to control the project budget and schedule (including contingencies).
- Institute contracting methods that select contractors who are committed to the goals of the project and the organization. Develop contract management procedures that hold contractors accountable for performance without creating a counterproductive adversarial atmosphere.
- Institute rigorous identification and control of changes, especially changes in scope. Make it clear that scope, budget, and schedule are inextricably linked and prohibit changes in scope that cannot be accommodated in the assigned budget.
- Provide consistent, uniform methods for tracking projects (e.g., earned value analysis) and disseminate this information so that all parties understand the status of every project with respect to its established scope, budget, and schedule.
- Provide a uniform financial accounting system for all projects.
- Train and qualify project managers in the classroom and on sites.
- Provide visible, recognized career paths for professional project managers.
- Assign increasing responsibilities to successful project managers.
- Create a climate of learning and openness to outside ideas, criticism, and standards through external project reviews, ISO 9000 certification, and participation in professional project and construction management organizations.
- Measure performance by results and provide positive incentives for the successful completion of projects on time and on budget.
- Provide a highly visible core competency in project management, an agent for cultural change, a role model, and a champion for project managers by establishing and supporting an office of project management that reports directly to the secretary. (NRC, 1999, pp. 75-76)

DOE has taken steps to implement many of these strategies to various degrees. Some, such as project manager training, are just beginning to be implemented and need time to demonstrate their effectiveness while others, such as providing a highly visible project management champion, have gone nowhere.

Nonetheless, the committee has observed signs of positive change in the DOE project management culture over the past 3 years. The development of a consistent set of project management principles and requirements and the recognition of project management as a discipline are providing a foundation on which

DOE management can build cultural change. Evidence that the highest levels of management are engaged in process improvement has been crucial in beginning this change and is also crucial to the institutionalization of an effective project management system. The committee met with many DOE personnel in the field and observed a good deal of energy focused on project management improvement. Although pockets of denial and resistance remain, the committee was impressed with the enthusiasm expressed at several field organizations. The committee's visits to the Hanford, Savannah River, Sandia, and Los Alamos complexes over the past year provided some encouraging signs of progress in improving project management competency among the federal employees at those sites. Particularly noteworthy was the growth in size from 4 to 20 project direction staff, in part through the use of interns, and the increased competency of the project direction staff at the Los Alamos Site Office over the past 2 years.

Examples of DOE personnel seeking and using lessons learned from completed and ongoing projects were evident in project briefings to the committee. At Hanford, a subcontractor had been engaged to incorporate operability and maintainability lessons learned from vitrification plants at Savannah River and West Valley into the Hanford waste treatment facility. Another, more general sign of positive change is the level of preparedness and knowledge of project details exhibited by project personnel at Hanford, Savannah River, Sandia, and Los Alamos.

## PROGRAM AND PROJECT PLANNING

### Strategic Planning

In its 2002 assessment report, the committee stressed the urgent need for the development of a strategic planning process to integrate program and project planning (NRC, 2003). The committee has observed significant progress in the development of these processes.

NNSA has recognized the situation and has taken active steps to reengineer the NNSA ESAAB process to integrate its portfolio management system with construction project execution and project outcomes. NNSA is commended for the development of its Future Years' Nuclear Security Plan (FYNSP), Ten Year Comprehensive Site Plans (TYCSPs), Integrated Construction Program Plan (ICPP), and 5-year budget process. The NNSA administrator has expressed his full support for these improvements in the project selection and execution process.

EM has begun its accelerated, 10-year closure planning process (a comprehensive new approach to project planning and management derived from the top-to-bottom review), with the full support and encouragement of the assistant secretary and undersecretary. The new closure plan could have a significant impact on the time and costs of site cleanup and closure. EM is to be commended for taking this comprehensive step toward gaining control of its site remediation projects for the benefit of the public, the environment, and taxpayers.

SC has begun to integrate its infrastructure needs beyond just the current budget period. These planning tools should continue to be developed and matured, and their effective use by management should improve project planning and project outcomes.

### Front-End Project Planning

The committee emphasized the need to perform effective front-end planning in both the 2001 and 2002 assessment reports (NRC, 2001b, 2003). The committee is encouraged by the improvements seen in front-end planning, especially the definition of project need at CD-0 and the identification of system requirements and alternatives at CD-1, on projects reviewed during site visits in the past year. Many mission need statements and acquisition planning documents, as well as early risk identification and management plans, show an improved emphasis on front-end planning.

The committee's assessment of the state of DOE front-end planning must also take into consideration the inconsistency of quality and incompleteness of project planning documents that the committee reviewed in the past year. The inconsistency of front-end planning is illustrated by the Glass Storage Facility at the Savannah River Site (SRS), despite the fact that it was essentially a replication of a previous project, which should have simplified the planning effort. A broad range of front-end planning issues is also highlighted in key findings of recent EIR reports (see Appendix F). As noted in the 2002 assessment report, consistency in the approach to and implementation of front-end planning is critical to overall project performance at DOE. Inconsistency is a symptom of problems with many issues, as outlined elsewhere in this report, including personnel issues, lack of training, limited accountability, and an immature project management culture.

### Project Budgets

As recommended in the 1999 NRC report, DOE has implemented procedures to avoid establishing project baselines before substantial engineering has been completed, in order to support the preparation of credible estimates. Prior to the availability of engineering data at CD-2, an estimated cost range is used to quantify expected project costs. However, DOE field personnel told the committee that internal pressure still exists to use the low end of the estimated cost range as the baseline cost. If this culturally driven practice continues, it will nullify the intent of the improved planning practices; prevent effective risk management; be inconsistent with *Guidelines and Discount Rates for Benefit-Cost Analysis of Federal Programs* (OMB Circular A-94); and demonstrate the persistence of bad practices, such as that referred to by GAO's 1998 criticism of DOE's habit of presenting rough order-of-magnitude numbers as point estimates (GAO, 1998). Using the cost estimate range until sufficient engineering data are available to support credible cost estimates is critical to improving project management and DOE's credibility with Congress.

## RISK MANAGEMENT

The committee has observed some improvement in DOE's risk management efforts, from the minimal and weak planning and mitigation observed in 2000, to documented planning for project risks and more active mitigation of risks during project execution. Improved risk management plans were reviewed by the committee at Hanford, Savannah River, Sandia, and Los Alamos. Risk mitigation at Savannah River has included risk monitoring and ongoing attention and documentation with tools such as risk registries. The Plutonium Packaging and Stabilization Project at Savannah River has integrated risk issues as a part of the project schedule. The committee continues to stress that improvements in risk management practices are essential to improving DOE project management (NRC, 2001b). This limited sample indicates a potentially significant change and demonstrates that good risk management practices do exist within DOE and could become widespread across the department if they were supported.

Although the committee found examples of improved risk management, the EIR summaries (see Appendix F) indicate that 8 of the 19 projects reviewed had key findings related to deficiencies in risk management or mitigation. Reviewing these and other projects during the past year, the committee believes that risk identification is not done consistently; active risk mitigation plans are not being developed and applied on many projects. The knowledge, skills, and abilities needed to perform and oversee risk management are inadequate, isolated, and not readily transferred to projects needing them because of the separate PSO structures and the reluctance to transfer human resources across the PSO boundaries. Risk management is an integral requirement of Order O 413.3 but has not been adequately addressed in Manual M 413.3-1 and in *Project Management Principles*.

DOE's projects are often first-of-a-kind projects that require extraordinary effort and the use of best practices and skills to manage risks. The committee believes that no project should pass CD-1 or CD-2 without an effective risk mitigation plan, especially the complex first-of-a-kind projects. A project with no risk management plan or an incomplete plan is not in control. A consistent approach to risk identification, assessment, and mitigation would be a first step toward making risk management more useful and usable across the department. The committee believes that DOE should develop such guidance and require its implementation for all projects.

## PROJECT CONTROLS

Project controls include earned value management systems (EVMSs); the Project Assessment and Reporting System (PARS); other reporting, change control, and change tracking systems; and other established management procedures. The ESAAB critical decision reviews and decision process are also considered to be project controls. The committee observed that most federal project directors now freely acknowledge the value of procedures such as the critical decision reviews and the related project justification and control activities. The committee believes that compliance with the requirements of Order O 413.3 corresponds to increased efforts to apply project controls for improved project performance. The committee also believes that DOE has made progress in some areas of project controls, but many others need additional improvement and more consistent application to be effective.

### Project Reporting

The Project Assessment and Reporting System is a Web-based distributed database for collecting and analyzing current project earned value data. The PARS manual notes that "the purpose of the DOE project reporting system is to deliver project status and assessment information to DOE senior managers and key program stakeholders" (DOE, 2003g, p. 1). The committee notes that PARS potentially could become a valuable tool for project monitoring, reporting, and oversight, although substantial improvements are needed to make it effective. PARS is in its second generation, but it has not yet been adequately refined to accomplish its objective. Further refinement of definitions of data-entry fields is needed to ensure the collection of consistent data. This problem is particularly troubling because typical PARS earned value data are 3 or more months out of date and PARS does not display the future project plan for budgeted cost of work scheduled (BCWS). Data-entry requirements should include planned budgets and schedules (e.g., BCWS) through the completion of the project, as well as historical performance data.

OECM has issued an updated users manual for PARS, but it gives no guidance on how to use or interpret the data. More follow-up is needed to determine who is using PARS, why they use it, and how it might be improved to increase its value to more users. The committee believes that a robust project database would be of considerable value as a project management tool, and it encourages DOE to continue to improve PARS and to extend its capabilities.

The use of PARS for any project management function is currently limited by its deficiencies. It cannot be used to assess the performance of projects in real time because of the lateness of data reporting (generally at least three months behind and often much more), nor can it be used to assess historical performance because an historical record of baseline changes is not provided. Owing to the failure to show baseline plans extended into the future (i.e., BCWS) and the lack of controls over changes to the baseline, evaluative indices of progress (cost performance index [CPI] and schedule performance index [SPI]) are so consistent that they are unbelievable to anyone with project management experience. Because of the incredible number of projects that are shown to be not approximately but *exactly* on schedule and on budget (i.e., $0.99 \leq SPI \leq 1.01$ and $0.99 \leq CPI \leq 1.01$), and considering other anomalies, the committee concludes that no confidence should be placed in PARS data and that PARS cannot be used to demonstrate whether or not DOE project management is effective or has even been improved. However, the committee believes that an effective project management culture would demand an effective project reporting system.

## Earned Value Management System

The committee recommended in previous reports that DOE apply a rigorous earned value management system and that it use this system to analyze and improve project performance (NRC, 1999, 2001b). DOE Order O 413.3 requires the implementation of EVMS on all projects over $20 million TPC (DOE, 2000b). The committee saw evidence in project management awards documents and in briefings to the committee that EVMS is being or will be used on projects below $20 million TPC. The application of EVMS is necessary for all projects that are monitored through the PARS database. From an examination of PARS data, the committee is concerned that the quality of EVMS department-wide is inconsistent.

OECM has initiated activities to provide EVMS training to DOE project directors and has proposed contracting with the Defense Contract Management Agency to verify contractor application of EVMS standards specified in O 413.3. EVMS is a critical part of project controls, and at this time there is no way of knowing if the reported data are accurate and reliable. The lack of evidence of consistent work breakdown structures, resource-loaded schedules, and baseline performance plans for cost and schedule causes the committee to doubt that EVMS has been effectively implemented.

## PERFORMANCE MEASURES

### Assessment Metrics

DOE does not have a uniform set of objective measures for assessing the quality of project management. The lack of objective measures or even reliable historic project data makes it difficult to assess progress in improving project management. It also makes it difficult to build confidence within GAO, Congress, OMB, and the public in the department's ability to manage the money it spends on its projects. Evidence continues to be anecdotal rather than objective, quantitative, and verifiable. The absence of objective performance measures prevents the identification of best practices and impedes widespread improvement in project management throughout the agency.

DOE Undersecretary Robert Card delegated acquisition executive authority to the assistant secretary for environmental management and the director of the Office of Science with the provision that their respective project management systems must be approved within a year (see Appendix E). However, the committee is not aware of any established DOE criteria for evaluating project management success, performance, or maturity. The Project Management Institute is developing its Organizational Project Management Maturity Model (OPM3) to assess the capabilities of an organization's procedures and personnel, but the model will not be ready until the beginning of 2004 at the earliest and would require some testing and validation before it could be used to evaluate DOE programs.

The committee has identified several lines of inquiry but does not have sufficient data to begin to discuss performance criteria. These lines of inquiry include the following:

- The percentage of DOE project managers who are certified professional project managers;
- The numbers of PMCDP courses offered and of personnel trained;
- The number of project directors and support staff for the value of projects managed;
- The functional quality of Integrated Project Teams;
- The quality of project planning documents (acquisition plans, project execution plans, risk management plans, and so on) submitted for ESAAB review;
- Management participation in ESAABs related to acquisition executive decisions;
- Effective use of performance-based contracts;
- Effective use of value engineering;
- Effective use of project controls;
- Trends in findings and comparison of independent project reviews and external independent reviews;
- Ratings by the EM Project Definition Rating Index (EM PDRI); and
- Comparison of actual project performance to original baselines for budget, schedule, and scope.

The committee believes that the approval of the program offices' project management systems should be based on a valid, comprehensive assessment. If such an assessment is not available, the committee recommends that the delegation of acquisition executive authority to the PSOs be revoked or continue to be provisional until additional direction on performance metrics and a revised assessment schedule can be provided and assessed.

The committee noted above that current project performance data available in PARS are not useful for assessing DOE project management because of problems with inconsistent data and the lack of historical trend data. Despite the recommendation in the 1999 NRC report (NRC, 1999) that DOE should develop a reliable database of all of its projects over time—past, present, and forthcoming—in order to be able to assess progress in improving project management, there is none 4 years later, and the committee has seen little evidence of progress in this regard.

The Construction Industry Institute's (CII's) Project Definition Rating Index was described in the 2001 assessment report as one example of an approach to assessing the maturity of front-end planning (NRC, 2001b). EM's adaptation of the PDRI continues to be used as a project review tool, but the committee has seen no evidence that the PDRI or any other approach has been used to calibrate DOE's front-end planning procedures (DOE, 2001a).

In the absence of a dedicated retrospective database of past DOE projects for comparison purposes, the capture of prospective data through PARS might be a way to eventually build a project database. To make this database useful for project analysis, PARS should archive the original cost and schedule baselines and all modifications to these baselines—constant rebaselining of projects precludes the ability to make valid comparisons over time and across projects. Although current revised baselines should be used to make day-to-day project management decisions, the original approved baselines and all subsequent changes should be tracked in order to evaluate overall project management performance.

By maintaining all baseline data, project management performance can be objectively assessed by comparing the actual total project costs with original budget estimates, actual completion dates with original deadlines, and delivered scope and quality with original project specifications. The variance between the original baselines and the cost, schedule, and scope at completion is one indicator of performance. Process improvement can be assessed by analysis of the trend of variances for projects initiated over a period of time. Unfortunately, this approach will require years before it produces usable results, but it is essential to long-term performance, and it should start now.

## Benchmarking

Benchmarking performance and management processes throughout a project's life cycle and from different perspectives can provide a basis for a measure of improvement of project management procedures. Both internal and external benchmarking perspectives are useful and should be a regular part of DOE benchmarking procedures. Internal benchmarking can compare projects across programs and across time. External benchmarking can compare DOE projects with those of similar complexity, size, and other characteristics in other federal agencies and in industry.

The committee found little evidence of external benchmarking (other than attempts by NA-54 to use CII benchmarking data), nor did it find evidence of internal benchmarking. The committee recognizes the inherent difficulty in benchmarking first-of-a-kind and one-of-a-kind projects but reiterates its belief that a consistent, continuously applied benchmarking program would be an effective tool for improving project management. The committee noted in its 2001 assessment that the PARS database should provide accurate, up-to-date information for use in benchmarking (NRC, 2001b), but there is no evidence that PARS data are used for this purpose.

## PROJECT REVIEWS

Prior to the 1998 NRC report, independent reviews in DOE were largely confined to those conducted by the Office of Science and mainly addressed the technical aspects of projects (NRC, 1998). Although these reviews neglected some aspects fundamental to sound project management, their contribution to executing technologically complex, one-of-a-kind projects is not to be minimized. OECM has issued a draft *Independent Review Procedures,* which outlines the departmental process and requirements (DOE, 2001b). The three major program offices (NNSA, EM, and SC) have each institutionalized and formalized their respective procedures in documents that cover the details and peculiarities pertinent to their programs (DOE, 2000c, 2000d, 2001a). These documents were reviewed in the 2001 assessment; the committee is not aware of any revisions since that time.

The procedures for the congressionally mandated EIRs have matured over time and the reviews are more comprehensive and have improved in quality. Consequently, there is increased appreciation within DOE of the value of EIRs, and DOE project directors interviewed by the committee now acknowledge that EIRs have provided useful information and have added considerable value to the project delivery process. Also, there has been more concentration on conducting rigorous reviews in the early planning stages. Because project cost and schedule are particularly difficult to evaluate in the early stages of a project, it is essential that well-qualified individuals or contractors perform these reviews. The committee has also observed value in DOE internal reviews, particularly those employing nonadvocate participants from across organizational lines.

### Criteria for Reviews

Although many DOE managers recognize the value of EIRs, the committee has heard the opinion voiced by some DOE managers and M&O contractors that external independent reviews provide no added value on projects costing $5 million to $20 million TPC. One PSO manager advocates that EIRs should not be required for projects costing less than $100 million TPC. The committee does not share this opinion. Review of the key EIR findings (see Appendix F) indicates persistent problems in planning and other areas across all Program Secretarial Offices for projects with less than $100 million TPC. On this basis, the committee concludes that there is no justification for reducing EIR requirements, at least until there is demonstrated improvement in project outcomes.

The committee continues to recommend that reviews of all projects should be conducted in some form. The scope and content of the review should be tailored to the complexities and peculiarities of the project, especially for those projects between $5 million and $20 million. By tailoring reviews to the complexity of projects, the expenditure of effort can be made compatible with the value added. Procedures already exist for waiving independent reviews for routine projects when this is justified, and there are procedures for tailoring reviews (DOE, 2001b).

Independent reviews of plans, assumptions, designs, estimates, and schedules are the accepted standard in industry. In fact, independent reviews are one of the means by which industrial firms become successful at project management and continue to stay that way (NRC, 2002b). The view that DOE should not have to do the things typically done by the most successful industrial owner firms does not withstand scrutiny. The committee strongly advocates that DOE continue to recognize the value of EIRs for improving project performance and expand their application for documenting lessons learned.

## Capability of Independent Reviewers

DOE has awarded three Management, Organization and Business Improvements (MOBIs) contracts for the performance of EIRs department-wide. The committee recognizes the benefits of such contracts for this type of work but believes that DOE needs to ensure that the contractors assign qualified personnel to each task. It is incumbent on DOE to exercise the necessary oversight to ensure the quality of reviews and to ensure that reviewers have the necessary qualifications and experience to add value to a project.

## Independent Cost Estimates

The committee reviewed independent cost estimates (ICEs) for a variety of projects. It found them to be essentially arithmetical checks of the extant estimate rather than critical analyses of the work breakdown structures, cost elements, risk assessments, and other factors that affect the accuracy and credibility of the estimates. The committee learned that the scope and definition of the cost estimate reviews have been changed to an independent cost review (ICR), which includes the needed critical analysis. If properly structured and performed, the ICRs may provide the rigorous analysis that the committee believes is lacking in the ICEs. However, there are no data for evaluating ICRs at this time.

## ACQUISITION AND CONTRACTING

In each of its previous reports, the committee stressed the linkage between effective acquisition and contracting, and successful project management (NRC, 2001a, 2001b, 2002a, 2003). These reports identified and encouraged the use of a variety of acquisition best practices, many of which have been described in the new manual, M 413.3-1 (DOE, 2003a).

For example, the manual requires that IPTs develop sound acquisition strategies to assess both risks and potential alternative contracting approaches. The OECM staff then assesses the completeness of the acquisition execution plans (AEPs) before forwarding them for senior management approval as part of the ESAABs and the critical decision process. This process is an essential means for senior managers to assess progress at critical steps.

The committee was pleased to see an example of the effective use of the acquisition process, when NNSA recently sought to develop a quick-response approach for assisting the Russian Federation in closing down its three remaining plutonium reactors and replacing them with fossil fuel plants. NNSA worked effectively to develop a reasonable comprehensive strategy and risk mitigation plan, consistent with OECM guidance. Moreover, the deputy secretary showed his commitment to the process, maintaining continuing control through at least the CD-1 milestone.

Other acquisition-related improvements include the following:

- *EM's development of a new contracting strategy relying on incentives for site closure.* EM is now applying that strategy to the Rocky Flats site.
- *Increased use of performance objectives in determining fees for site contractors.* DOE increased the proportion of contractor fees tied to performance objectives from 34 percent in FY 1996 to 70 percent in FY 2001 (GAO, 2003).
- *Increased use of competition among contractors to achieve best value for the department.* Of 16 Federally Funded Research and Development Center (FFRDC) contracts, 6 have been competed, and other competitions are planned. Moreover, the percentage of major site contracts awarded competitively increased from 38 percent in FY 1996 to 56 percent in FY 2001 (GAO, 2003).
- *The effective use of alternative contracting techniques to meet DOE needs.* An example is Los Alamos National Laboratory's use of a design-build approach for one of its major construction projects.

These examples demonstrate good front-end planning, senior management involvement, the effective use of performance metrics and incentives, and flexibility in contracting approaches. The committee strongly supports continuing the emphasis on all of these techniques.

In its previous reports, the committee stressed the benefits of systematic training in the use of performance-based contracting methods and encouraged the department to collect data on the effectiveness of these techniques. It reiterates the need to follow through on these recommendations.

## RECOMMENDATIONS

### Conclusion

DOE has made significant progress in improving project management through organizational changes and the development of policies and procedures. These changes were completed in 2003 with the release of the manual *Project Management for the Acquisition of Capital Assets* (Manual M 413.3-1) and have just begun to be implemented. The committee believes that current policies and procedures need to be fully and consistently implemented, and that opportunities exist for additional changes that are needed in order to maintain the progress achieved to date and to continue the improvement process so as to bring DOE's project management capabilities to an appropriate level of excellence. The committee believes that action on most of its past recommendations is still incomplete and that the results department-wide are inconsistent. The opportunities for continued improvement in the application of accepted project management practices are presented above in this chapter and are summarized below as committee recommendations.

### Recommendations

- DOE senior managers should actively use the Office of Program Analysis and Evaluation to provide objective analysis and advice on mission need and project cost-benefit justification at the early stages of incipient projects and should use the Office of Engineering and Construction Management to standardize requirements for management decisions, monitor the progress of projects, and provide objective analysis and advice. These offices, as well as the project management support offices and the DOE site offices, should be provided adequate resources, staffing, training, and moral support.
- The DOE deputy secretary either should be the champion for project management improvement—to develop project management into a core competency of the department, to assure that the department maintains an adequate staff of qualified project directors to manage its portfolio of projects, and to assure that the disciplined execution of projects is a priority for managers at all levels—or should appoint someone to perform this role, reporting to the deputy secretary (see footnote on page 16).
- Just as the deputy secretary should be a champion for project management

throughout the department, the program heads should be strong, visible champions for improved project management in their programs (see footnote on page 16).

- Current policies and procedures should be kept intact and revised only as necessary to improve the planning and execution of projects. The applicability of project management policies and procedures should remain at current levels. Requirements should be tailored to the size and complexity of projects, but exemptions for projects or sites should not be considered until such decisions can be supported by a record of excellence in project management and project performance is established.
- Manual M 413.3-1 should define the roles and responsibilities required by Order O 413.3. It should cover the DOE chain of responsibilities from the acquisition executive to the project director, and the contractor chain of responsibilities from executives to managers in the field. The definitions should describe responsibilities to develop, review, comment, approve, and execute at each step of the DOE capital acquisition process. In addition, each of the Program Secretarial Offices needs to develop a detailed "roles and responsibilities" document to reflect the different procedures used by their separate organizations.
- If the application of policies and procedures results in delays because resources to perform reviews expeditiously are lacking, then more resources should be provided. Steps should also be taken to streamline the critical decision review process by eliminating unnecessary sign-offs.
- DOE should ensure that funding and implementation of the Project Management Career Development Program are priorities. The committee sees particularly urgent needs for department-wide training in front-end planning, risk analysis, and project controls. Training for project directors and project support staff should be centrally funded and should utilize alternatives to traditional classroom delivery of instructional materials.
- DOE should adopt a department-wide strategy to develop a sustainable, qualified workforce for directing the projects required to achieve the department's missions. DOE should work with Congress to allow funding for project direction to be included in project budgets.
- DOE should provide contracting officer representative authority to project directors.
- DOE should continue and expand department-wide and program-specific project management workshops to recognize project management achievements, reinforce the professional identity of project directors, and share lessons learned.
- Strategic planning initiatives should be continued, expanded, and used by management to improve project selection, planning, and execution.
- DOE should develop detailed procedures and guidance for identifying risks, planning strategies to address risks, and managing risks throughout

the life cycle of projects, and should require their implementation for all projects. Projects should not pass CD-1 or CD-2 without an effective risk mitigation plan.

- Manual M 413.3-1 and *Project Management Practices* should be expanded to include detailed procedures and deliverables required between CD-1 and CD-2. Federal project directors should be involved in this development effort so as to incorporate the best industry practices in DOE procedures.

- Steps should be taken to ensure the quality and consistent application of project controls. Planned project scope, budgets, and schedules should be maintained through completion of a project. Although current revised baselines should be used to make day-to-day project management decisions, the original approved baselines and all subsequent changes should be retained in order to evaluate overall project management performance.

- Steps should be taken to ensure that the Project Assessment and Reporting System (PARS) data are timely, accurate, and consistent from project to project. Analytical tools and presentations should be enhanced. PARS should archive the original cost and schedule baselines and all modifications to these baselines.

- DOE should develop consistent performance metrics for evaluating project management maturity department-wide. Both internal and external benchmarking should be a regular part of procedures for all phases of projects.

- DOE should continue to recognize the value of external independent reviews for improving project performance and should expand their application for documenting lessons learned.

- DOE should exercise the necessary oversight to ensure the quality of reviews and to ensure that reviewers have the necessary qualifications and experience to add value to the project.

- DOE should follow previous recommendations to provide systematic training in the use of performance-based contracting methods and should collect data on the effectiveness of these techniques.

## REFERENCES

Collins, James C. 1994. Built to Last. New York, N.Y.: HarperCollins.

DOE (Department of Energy). 1999. Memorandum for All Departments from T.J. Glauthier, Deputy Secretary; Subject: Project Management Reform Initiative. June 25.

DOE. 2000a. Program and Project Management Policy for the Planning, Programming, and Acquisition of Capital Assets (Policy P 413.1). Washington, D.C.: Department of Energy.

DOE. 2000b. Program and Project Management for the Acquisition of Capital Assets (Order O 413.3). Washington, D.C.: Department of Energy.

DOE. 2000c. Office of Defense Programs Project Review Procedures. Washington, D.C.: Department of Energy.

DOE. 2000d. Office of Environmental Management Internal Independent Review Handbook. Washington, D.C.: Department of Energy.

DOE. 2001a. Office of Science Independent Review Handbook. Washington, D.C.: Department of Energy.

DOE. 2001b. Office of Engineering and Construction Management Draft Independent Review Procedures. Washington, D.C.: Department of Energy.

DOE. 2002. Value Engineering (Notice N 413.2). Washington, D.C.: Department of Energy.

DOE. 2003a. Project Management for the Acquisition of Capital Assets (Manual M 413.3-1). Washington, D.C.: Department of Energy.

DOE. 2003b. Acquisition Career Development Program, Attachment 4 (Order O 361.1). Washington, D.C.: Department of Energy.

DOE. 2003c. Office of Management and Budget Evaluation. Available online at http://www.mbe.doe.gov/crOrg/me20.htm. Accessed September 9, 2003.

DOE. 2003d. Office of Engineering and Construction Management. Available online at http://oecm.energy.gov/. Accessed September 9, 2003.

DOE. 2003e. Value Engineering (Acquisition Letter 2003-4). Washington, D.C.: Department of Energy.

DOE. 2003f. Project Management Practices. Available online at http://oecm.energy.gov/. Accessed September 9, 2003.

DOE. 2003g. Project Assessment and Reporting System User Manual, version 3.0. Washington, D.C.: Department of Energy.

GAO (General Accounting Office). 1998. Nuclear Waste: Department of Energy's Hanford Tank Waste Project: Schedule, Cost, and Management Issues (GAO/RCED-9913). Washington, D.C.: General Accounting Office.

GAO. 2003. Department of Energy: Status of Contract and Project Management Reforms (GAO-03-570T). Washington, D.C.: General Accounting Office.

NRC (National Research Council). 1998. Assessing the Need for Independent Project Reviews in the Department of Energy. Washington, D.C.: National Academy Press.

NRC. 1999. Improving Project Management in the Department of Energy. Washington, D.C.: National Academy Press.

NRC. 2001a. Improved Project Management in the Department of Energy. Letter report, January. Washington, D.C.: National Academy Press.

NRC. 2001b. Progress in Improving Project Management at the Department of Energy, 2001 Assessment. Washington, D.C.: National Academy Press.

NRC. 2002a. Progress in Improving Project Management at the Department of Energy, 2002 Interim Assessment. Letter report, May. Washington, D.C.: National Academy Press.

NRC. 2002b. Proceedings of Government/Industry Forum: The Owner's Role in Project Management and Preproject Planning. Washington, D.C.: National Academy Press.

NRC. 2003. Progress in Improving Project Management at the Department of Energy, 2002 Assessment. Washington, D.C.: The National Academies Press.

U.S. Senate. 2003. The Energy Policy Act of 2003 Report of the Committee on Energy and Natural Resources, United States Senate, to Accompany S. 1005 (Report 108-43). Washington, D.C.: U.S. Government Printing Office.

# 3

# Prognosis for Progress

## INTRODUCTION

In its 3 years of existence, this committee has observed progress in the improvement of project management at the U.S. Department of Energy. During this time, the committee has also identified issues that need continuing attention if DOE is to achieve the level of competence in project management required to fulfill its missions. However, the committee is concerned that the rate of improvement may be too slow and that the momentum for continued improvement of project management may not be sustained. This concern is based on three critical factors: (1) the absence in DOE of a recognized champion for project managers and process improvement—an individual who is at a level of authority to ensure adherence to policies and procedures and the availability of the necessary funding and personnel resources,[1] (2) inconsistent project performance, and (3) the slow rate of acceptance and continued pockets of resistance to project management reforms.

---

[1]Editor's note: The 2003 assessment is based on information reviewed by the committee through September 2003. In December 2003 the DOE deputy secretary appointed an associate deputy secretary with responsibilities for capital acquisition and project management, a positive step of which readers should be aware as they consider the committee's comments and recommendations regarding the need for a strong and visible champion of project management issues in DOE.

## THE PROSPECT FOR CONTINUED IMPROVEMENT

A delegation of the committee met in June 2003 with the heads of the major DOE programs—the Office of Environmental Management (EM), the National Nuclear Security Administration (NNSA), and the Office of Science (SC)—to discuss the status of project management and the prospect for continued improvement in their respective programs. A brief discussion of each meeting follows.

### Office of Environmental Management

The assistant secretary noted that EM is in the business of solving problems and that undertaking the right projects depends on establishing an accurate problem definition before defining a solution. These tasks are different from the tasks required to plan typical capital acquisition projects, but the procedures defined by Order O 413.3 are applicable. She also noted that tailoring project requirements and delegation of approval authority in EM will be dependent on demonstrated competency in planning and executing projects. EM supports the development and coordination of a professional training and career development program through a departmental structure.

### National Nuclear Security Administration

The administrator noted that NNSA has done much more to improve project management than the committee gave credit for in its 2002 assessment report. The administrator offered the perspective that NNSA has limited resources to complete its mission and needs to avoid duplicating efforts among DOE headquarters, the field, and contractors. NNSA is challenged to get the right people in the right place to undertake a disciplined process of project management. The development of a lessons-learned database is being discussed as a means to benefit from past problems and success. The recovery of the National Ignition Facility (NIF) from its project management problems was cited as one of the successes, which illustrates that the agency can learn to improve project management. The committee noted that one of the major reasons for NNSA's difficulties in improving project management is the inadequate size and professionalism of the project management staff. The committee also noted that the administrator needs to provide highly visible attention to project performance in order to accomplish these objectives: to show that excellence in project management is expected and important, to communicate his requirements and expectations to DOE personnel and contractors, to hold managers consistently accountable for project performance, to use the advice of expert staff, and, most importantly, to show that competent project management is a priority by providing the resources needed to manage projects and improve the project management process.

**Office of Science**

The director of the Office of Science took strong exception to aspects of the committee's 2002 assessment of project management at DOE. (See Appendix G for relevant correspondence.)

The positions expressed by the Office of Science during the June 2003 meeting are that O 413.3 should not apply to projects with less than $100 million total estimated cost (TEC), that external independent reviews (EIRs) add no value, that project manager professional certification is not valuable, and that the Project Management Career Development Program (PMCDP) is too prescriptive. The director noted that SC is considered by the international physics community to be a leader in project management.

On the basis of the factual evidence with respect to ongoing problem areas (see Appendix F for a compilation of key EIR findings), the committee finds no support for the contention that projects under $100 million TEC should be exempted from Order O 413.3 reviews or from the requirement for EIRs. And, as noted in previous assessments and reiterated in this one, the committee continues to support project manager professional certification and the PMCDP.

The committee recognizes, and has stated in a number of places, that the highly technical, first-of-a-kind projects undertaken by SC and NNSA require an approach to project management that is different from the approach taken for more routine infrastructure projects; however, the committee believes that DOE's current policies and procedures define a minimum level of detail needed by DOE for effective project management, and that they should be applied to all projects. The committee also believes that the current provisions for tailoring requirements are sufficient to assure that management procedures are cost-effective. The committee noted in its 2001 assessment that:

> The use of techniques and skills that are appropriate to low-uncertainty projects can lead to poor results when applied to high-uncertainty projects with great potential for changes and high sensitivity to correct decisions. For high-risk projects, a flexible decision-making approach is much more successful. (NRC, 2001b, p. 41)

The committee discussed first-of-a-kind and science projects extensively in previous assessments. For example, the 2002 assessment report (NRC, 2003) devotes an entire chapter (Chapter 7, pp. 40-48) and an appendix (Appendix F, pp. 108-111) to this subject. The committee noted in its 2002 assessment report that:

> First-of-a-kind projects have been and can be successfully managed and execut-ed by DOE, but they require particular care. The higher degree of uncertainty that attends these projects requires managers who are experienced in dealing with uncertainty and ambiguity. Not all project managers have this ability. The

best project managers and management systems more than pay for themselves on first-of-a-kind projects by delivering projects on schedule with little budget overrun. (NRC, 2003, p. 48)

The committee provided the following finding and recommendation regarding the management of first-of-a-kind projects:

**Finding**: Innovative, cutting-edge, and exceptional risk management abilities are needed by DOE to identify and address the risks in many of its projects. DOE needs to develop expertise and excellence in managing very risky development projects. The DOE complex has the intellectual, computational, and other resources necessary to produce significant improvements in this area.

**Recommendation:** DOE should develop more expertise and improved tools for risk management. Nontraditional and innovative approaches, tools, and methods should be investigated for their adaptability to DOE project conditions and use in DOE risk management. (NRC, 2003, pp. 47-48)

The committee believes that SC has the capacity to perform excellent project management for all of its projects, but it needs to first recognize the impediments and take action to remove them. The SC director stated in a May 23, 2002, policy memorandum (DOE, 2002) that project management should have the following objectives:

1. Ensure that projects clearly support program research missions and strategic plans in a cost-effective manner,
2. Verify that projects are adequately defined and staffed before committing significant resources,
3. Establish a project baseline in terms of scope, schedule, and cost,
4. Maintain the project baseline through formal change control, and
5. Determine a project's success by measuring performance against the approved baseline. (DOE, 2002, p. 1)

The *Project Management Improvements Committee Report* attached to the memorandum outlined the actions that SC would take to ensure these objectives are met. The committee applauds this direction and notes that it applies to all projects covered by O 413.3 and should, if fully implemented, improve the management of SC projects.

## PROJECT PERFORMANCE

The 2002 and 2003 DOE Project Management Awards demonstrate the contribution of excellent front-end planning practices to successful projects and provide evidence that DOE can do projects well. The committee cited the 2002 award recipients in its 2002 assessment (NRC, 2003) and congratulates the following projects, which received awards in 2003:

- Fissile Materials Disposition, Highly Enriched Uranium Blend Down Project—Secretary's Excellence in Acquisition Award;
- Nonproliferation and International Security Center Project—Secretary's Award of Achievement and the Secretary's Acquisition Improvement Award;
- Rocky Flats Field Office, Building 371 Closure—Secretary's Award of Achievement;
- Rocky Flats Field Office, Building 771/774 Closure Project—honorable mention;
- Savannah River Operations Office, K Area Nuclear Material Storage Project—honorable mention;
- Oak Ridge Operations, EM, Hydrofracture Well Plugging and Abandonment Project—honorable mention;
- Chicago Operations Office, Tokamak Fusion Test Reactor Decontamination and Decommissioning Project—honorable mention;
- Oak Ridge Operations, EM, Facilities Revitalization Project—honorable mention; and
- Environmental Management Waste Management Facility Project—honorable mention.

However, not all DOE projects have performed to the level of those listed above. OECM recently compiled a list of significant adverse findings identified during EIRs performed in FY 2001 and FY 2002 prior to CD-2 baseline validations (see Appendix F). This compilation covers 19 projects with TECs under $100 million, equally apportioned between NNSA, EM, and SC. The purpose of the list is to detect the most pervasive and repetitive problems, as well as trends in problem correction.

The committee analyzed the compilation of key findings and notes a wide variance between projects in the number and the subject matter of the key findings. Only one project received a clean bill of health. EM projects, particularly at the Idaho National Engineering and Environmental Laboratory, and SC projects, particularly at Oak Ridge National Laboratory, exhibited a wide range of deficiencies. The NNSA projects had fewer deficiencies and appeared to be more attuned to the requirements of baseline validation.

In reviewing the list, the committee notes that the major problems are as follows: (1) inadequate project definition, (2) lack of documented rationale for decisions, (3) weak risk assessment and/or risk management plans, (4) haphazardly setting contingency allowances that are not necessarily based on risk, and (5) lack of integrated resource and cost-loaded schedules. These issues have been repeatedly addressed in previous committee reports. The committee concludes that apparently not all DOE field organizations are cognizant of or competent in the processes and procedures that should be undertaken between CD-1 and CD-2. This may be due in part to inadequately defined requirements for CD-2. The quality of project management remains inconsistent from project to project. Although some sites and some projects may perform better than others, the committee was chartered to assess project management across the entire department, not to identify pockets of superior performance. On the basis of the factual evidence it reviewed, the committee finds no basis to support the contention that specific program secretarial offices, sites, or management and organization contractors should be exempted from the requirements of O 413.3. The fact that performance varies across sites, laboratories, and programs is not considered a virtue; rather, it illustrates the ongoing need for establishing consistent policies and procedures, for transferring lessons learned, and for overcoming cultural resistance to communication and cooperation among the competing elements of the DOE complex. As the committee was told by a DOE employee at Oak Ridge, "We recognize no authority outside the Office of Science"—a revealing statement about the attitudes still prevalent in DOE.

These observations lead to the conclusion that DOE project management is not yet a process in full control. However, project management reforms initiated over the past 3 years have been successful in diminishing some of the differences across departmental program offices.

## MOMENTUM FOR IMPROVEMENT

The committee believes that since DOE launched its project management reform initiative in 1999, the department has made progress in improving management procedures and project performance. There are no objective performance measures in place by which to document progress, but this conclusion is substantiated by the committee's observations in this and preceding reports over a 3-year period. Today, the consensus of the committee is that DOE project management has significantly improved in the past 3 years but that the process is far from complete. The committee is concerned that the rate of progress may not be sustainable or fast enough to allow the department to achieve competence, let alone excellence, in project management.

The committee is aware of large private companies that have improved their project management process from poor to excellent in this same amount of time (NRC, 2002). The common factors that drove improvements in these companies are (1) a commitment from top management, (2) a strong, visible champion for project management and process improvement, and (3) a consistent, disciplined process with an emphasis on front-end planning. The case studies presented at the 2001 government/industry forum demonstrated that excellence in project management in industry is achieved only when the chief executive officer (CEO) or chief operating officer (COO) becomes convinced that excellence in project management is essential to the success of the corporate mission, puts the resources and prestige of his or her position behind it, appoints a project management champion reporting directly to the CEO or COO, and becomes directly involved in approvals of project plans from the earliest stages. There is no shortcut or secret method, and it is not glamorous. In these companies, commitment to the corporate position on project management becomes a condition of employment. The committee did not observe this consistent level of commitment throughout DOE.

The NRC 1999 report summarized the status quo for DOE project management as follows:

> The fundamental deficiency is DOE's organization and culture, which do not provide a focus for project management. As a result, the processes used by field offices, operations offices, and their contractors for planning and executing projects are inconsistent; lessons learned about cost estimating techniques, project review processes, change control mechanisms, and performance metrics are not transferred from one project to another; and there is no systematic program for recruiting and training professional project managers and no career path for project management. Related fundamental problems are a general lack of accountability and unclear lines of authority. (NRC, 1999, p. 2)

Unfortunately, despite significant progress in departmental project management policies and procedures, DOE-wide implementation of policies and procedures in 2003 can still be characterized as the old status quo.

The committee observed impediments to sustained, rapid improvement of project management at DOE, which are discussed in the previous chapters. These impediments all are correctable and could have been corrected long ago, so their persistence is an indication of more-deep-seated problems. Congress, DOE senior managers, DOE personnel at all levels, and DOE contractors need to affirm the critical importance of good project management to the success of the enterprise and devote the resources, both human and financial, to continuous improvement in project management. If not, other issues will continue to compete for senior manager and staff time and attention, and initiatives to improve project management may wither and ultimately die.

With the publication of project management policies and procedures (Order O 413.3, Manual M 413.3-1, *Project Management Practices*) and the establishment of the PMCDP, which must now be implemented and executed, DOE is at a crossroads in project management. If there is visible recognition by senior management of accomplishments and strong support for continued process improvement, DOE can ultimately achieve and sustain an acceptable level of competence in project management. However, if DOE senior management does not visibly demonstrate a continual interest in actively working to assure implementation of process improvements, DOE is likely to revert to the quality of project management observed in 1999.

DOE senior management has shown the ability to step in to resolve deadlocks, but this is short-term firefighting. As stressed by the administrator of NNSA, senior managers have many other demands on their time and cannot spend much of it on project management. Unfortunately, at this time the signals from DOE senior managers are mixed—this report and its predecessors have shown that DOE management has not fully committed to applying the resources necessary for success. The committee is convinced that permanent improvement in project management at DOE requires both continual leadership from senior management and a recognized project management champion with adequate authority who can and does spend full time on improving project management.

The committee has observed that permanent improvement in DOE project management is impeded by high personnel turnover, inadequate numbers of project personnel, and inadequate training of project management personnel. Since its January 2001 letter report (NRC, 2001a), the committee has urged DOE to institute a career development program to improve personnel retention, to expand the inadequate staff of professional project managers (project directors), and to institutionalize best practices by implementing policies and procedures. Because change in leadership at DOE is inevitable, the project management champion must strive to institutionalize improvements in the organization, policies, procedures, and project management culture throughout the department.

Therefore, the committee believes that it is critical to the continued improvement of project management in the department to have a project management champion with the authority to assure that the project management viewpoint is expressed in all decisions as well as to guide, support, and develop a professional project management staff across the department, with the ultimate goal of achieving and sustaining excellence in DOE project management (see footnote on page 41).

## FINDINGS AND RECOMMENDATIONS

**Finding.** DOE has recognized its fundamental deficiencies in project management and initiated reforms to improve procedures. DOE has made significant progress in issuing policies and procedures, but the implementation process is not complete. For example, Manual M 413.3-1, *Project Management for the Acquisition of Capital Assets*, was issued only in March 2003, and it has not yet been fully implemented. Substantial additional effort will be needed to create and sustain a culture that includes excellence in project management and project performance.

**Recommendation.** DOE should continue the course set by Policy P 413.1, Order O 413.3, and Manual M 413.3-1 to guide the planning and execution of projects. Senior managers should take visible and meaningful action to reaffirm the current direction and to assure that resources are provided for managers and their support staff at all levels so that they have the tools and knowledge needed to effectively implement policies and procedures.

**Finding.** The NRC report *Improving Project Management in the Department of Energy* included a set of findings and recommendations as a guide for improving project management at DOE (NRC, 1999). The 2001 and 2002 assessment reports provided additional guidance on specific aspects of project management that are critical to improving project performance (NRC, 2001b, 2003). (See Appendix D for a compilation of findings and recommendations.)

**Finding.** The committee has seen significant progress in the development of policies and procedures to implement the committee's prior recommendations; however, these policies and procedures have not yet been fully implemented. The committee has also identified specific areas that need additional improvement.

**Recommendation.** The committee continues to endorse the recommendations in its previous reports. DOE should continue to use these reports and recommendations as a guide to completing the changes needed to continue improving project management.

**Recommendation.** DOE's current policies and procedures with respect to project management should be maintained, and steps should be taken to improve them over time. Senior management should take visible actions to assure that project management policies and procedures are implemented correctly and consistently department-wide. Requirements should be tailored to the size and complexity of projects, but exemptions for projects or sites should not be considered until such decisions can be supported by a record of excellence in project management and until project performance is established.

**Finding.** The 1999 NRC report recommended that efforts to improve project management should be led by an office of project management and noted that, to be successful, it needed to be a top-down management initiative with the full support of managers at all levels of the department (NRC, 1999). The project management support offices, the Office of Engineering and Construction Management, and the Office of Program Analysis and Evaluation were created to fill this need and have produced significant accomplishments. Continued improvement will require strong-willed leadership, a focus on results, stability of processes and procedures, accountability with consequences, and a serious expenditure of resources to make the needed changes. The committee is convinced that permanent improvement in project management at DOE requires both continual leadership from senior management and a recognized project management champion with adequate authority who can and does work full-time on improving project management.

**Finding.** The committee has observed that permanent improvement in DOE project management is impeded by high personnel turnover, inadequate numbers of project personnel, and inadequate training of project management personnel. Since its January 2001 letter report (NRC, 2001a), the committee has urged DOE to institute a career development program to improve personnel retention, to expand the inadequate staff of professional project managers (project directors), and to institutionalize best practices by implementing policies and procedures. Because change in leadership at DOE is inevitable, the project management champion should strive to institutionalize improvements in the organization, policies, procedures, and project management culture throughout the department.

**Recommendation.** The DOE deputy secretary is the department's chief operating officer and chief acquisition executive. As such, the deputy secretary has been given the responsibility for assuring that projects are effectively planned and executed. To perform these functions, the deputy secretary either should be the champion for project management improvement or should appoint someone to perform this role, reporting to the deputy secretary (see footnote on page 41).

**Recommendation.** The DOE program heads for the Office of Environmental Management, the National Nuclear Security Administration, and the Office of Science have been delegated authority as acquisition executive for projects under $400 million and are responsible for project management and performance for their respective program offices. To perform these functions, the program heads should visibly and actively promote and defend efforts to improve project management capabilities and their consistent application throughout the department.

## REFERENCES

DOE (Department of Energy). 2002. Memorandum from Raymond Orbach, Subject: Office of Science Direction on Project Management. May 23.

NRC (National Research Council). 1999. Improving Project Management in the Department of Energy. Washington, D.C.: National Academy Press.

NRC. 2001a. Improved Project Management in the Department of Energy. Letter report, January. Washington, D.C.: National Academy Press.

NRC. 2001b. Progress in Improving Project Management at the Department of Energy, 2001 Assessment. Washington, D.C.: National Academy Press.

NRC. 2002. Proceedings of Government/Industry Forum: The Owner's Role in Project Management and Preproject Planning. Washington, D.C.: National Academy Press.

NRC. 2003. Progress in Improving Project Management at the Department of Energy, 2002 Assessment. Washington, D.C.: The National Academies Press.

# Appendixes

# APPENDIX A

# Statement of Task

In response to a congressional directive, the National Research Council has appointed a committee to review and assess the progress made by the U.S. Department of Energy (DOE) in improving its project management practices. This study includes evaluation of the implementation of recommendations in the 1999 NRC report *Improving Project Management in the Department of Energy.* The principal goal of this effort is to assess DOE's efforts to improve project management practices, including: (1) specific changes in organization, management practices, personnel training, and project reviews and reporting; (2) an assessment of the progress made in achieving improvement; and (3) the likelihood that improvements will be permanent. These tasks will also require development of a framework for evaluation and performance measures specifically tied to DOE's project management process.

# APPENDIX B

# Biographies of Committee Members

**Kenneth F. Reinschmidt** (National Academy of Engineering) is professor of civil engineering and holds the J.L. Frank/Marathon Ashland Petroleum LLC Chair in Engineering Project Management at Texas A&M University. He retired from Stone & Webster as senior vice president. He was appointed chair of this committee for his combination of expertise in the disciplines of civil engineering, project management, cost estimating, and the management of large-scale construction projects, including nuclear and fossil fuel power plant construction. He held various positions at Stone & Webster, including president and CEO of Stone & Webster Advanced Systems Development Services, Inc., and manager of the consulting group in the Engineering Department. In these positions he was engaged in structural engineering, operations research, cost analysis, construction engineering and management, and project management. Prior to his work at Stone & Webster, Dr. Reinschmidt was a senior research associate and associate professor in the Civil Engineering Department at the Massachusetts Institute of Technology, where he was engaged in interdisciplinary research on power plant engineering, design, construction, and project management. Dr. Reinschmidt served as chair of the committee that produced the recent NRC report *Improving Project Management in the Department of Energy* and was reviewer of the NRC report *Assessing the Need for Independent Project Reviews in the Department of Energy*. He is a former member of the Building Research Board of the National Research Council and served on or chaired several other NRC committees, including the Committee on Integrated Database Development, the Panel for Building Technology, the Committee on Advanced Technology for Building Design, and the Committee on Foam Plastic Structures. He has also served on

54

several National Science Foundation review panels on construction automation, computer-integrated construction, and engineering research centers. He obtained his B.S., M.S., and Ph.D. degrees from the Massachusetts Institute of Technology.

**Don Jeffrey (Jeff) Bostock** retired from Lockheed Martin Energy Systems, Inc., as vice president for engineering and construction with responsibility for all engineering activities at the Oak Ridge nuclear complex. He is serving on this committee because of his experience with managing projects as a DOE contractor. He has also served as vice president of defense and manufacturing and manager of the Oak Ridge Y-12 plant, a nuclear weapons fabrication and manufacturing facility. His career at Y-12 included engineering and managerial positions in all of the various manufacturing, assembly, security, and program management organizations. He also served as manager of the Paducah Gaseous Diffusion Plant, which provides uranium enrichment services. He was a member of the committees that produced the NRC reports *Proliferation Concerns: Assessing U.S. Efforts to Help Contain Nuclear and Other Dangerous Materials and Technologies in the Former Soviet Union* and *Protecting Nuclear Weapons Material in Russia*. Mr. Bostock also served as a panel member for the annual NRC assessment of the Measurement and Standards Laboratories of the National Institute of Standards and Technology. Mr. Bostock has a B.S. in industrial engineering from Pennsylvania State University and an M.S. in industrial management from the University of Tennessee. He is a graduate of the Pittsburgh Management Program for Executives.

**Donald A. Brand** (National Academy of Engineering ) retired from the Pacific Gas and Electric (PG&E) Company as senior vice president and general manager, engineering and construction business unit. He more recently was a lecturer at the University of California at Berkeley, teaching construction management. Mr. Brand was appointed as a member of this committee because of his expertise in the management of the design, engineering and construction of large, complex energy-related facilities. During his 33 years with PG&E, he carried out numerous managerial and engineering responsibilities related to the design, engineering, construction, and operation of fossil fuel, geothermal, nuclear, and hydroelectric generating facilities, as well as of electrical transmission, distribution, and power control facilities. Mr. Brand's industry activities have included membership on the Electric Power Research Institute's Research Advisory Committee and on the Association of Edison Illuminating Companies' Power Generation Committee. He has been a member of numerous NRC committees. He belongs to numerous professional societies and is a registered professional engineer in California. He received a B.S. in mechanical engineering and an M.S. in mechanical (nuclear) engineering from Stanford University. He also graduated from the Advanced Management Program of the Harvard University School of Business.

**Allan V. Burman** is president of Jefferson Solutions, a division of the Jefferson Consulting Group, a firm that provides change management services and acquisition reform training to many federal departments and agencies. He serves as a member of this committee because of his expertise in federal acquisition, procurement, and budget reform. Dr. Burman provides strategic consulting services to private sector firms doing business with the federal government as well as to federal agencies and other government entities. He also has advised firms, congressional committees, and federal and state agencies on a variety of management and acquisition reform matters. Prior to joining the Jefferson Consulting Group, Dr. Burman had a long career in the federal government, including serving as administrator for federal procurement policy in the Office of Management and Budget (OMB), where he testified before Congress over 40 times on management, acquisition, and budget matters. Dr. Burman also authored the 1991 policy letter that established performance-based contracting and greater reliance, where appropriate, on fixed-price contracting, as the favored approach for contract reform. As a member of the Senior Executive Service, Dr. Burman served as chief of the Air Force Branch in OMB's National Security Division and was the first OMB branch chief to receive a Presidential Rank Award. Dr. Burman is a fellow and member of the board of advisors of the National Contract Management Association, a principal of the Council for Excellence in Government, a director of the Procurement Round Table, and an honorary member of the National Defense Industrial Association. He is also a contributing editor and writer for *Government Executive* magazine. Dr. Burman obtained a B.A. from Wesleyan University; was a Fulbright Scholar at the Institute of Political Studies, University of Bordeaux, France; and has a graduate degree from Harvard University and a Ph.D. from the George Washington University.

**Lloyd A. Duscha** (National Academy of Engineering) retired from the U.S. Army Corps of Engineers in 1990 as the highest-ranking civilian after serving as deputy director, Engineering and Construction Directorate, at headquarters. He serves as a member of this committee because of his expertise in engineering and construction management and his roles as principal investigator for the NRC report *Assessing the Need for Independent Project Reviews in the Department of Energy* and member of the committee that produced the NRC report *Improving Project Management in the Department of Energy*. He served in numerous progressive Army Corps of Engineer positions in various locations over four decades. Mr. Duscha is currently an engineering consultant to various national and foreign government agencies, the World Bank, and private sector clients. He has served on numerous NRC committees and recently served on the Committee on the Outsourcing of the Management of Planning, Design, and Construction Related Services as well as the Committee on Shore Installation Readiness and Management. He chaired the NRC Committee on Research Needs for Transuranic and Mixed Waste at Department of Energy Sites and serves on the Committee on

Opportunities for Accelerating the Characterization and Treatment of Nuclear Waste. He has also served on the Board on Infrastructure and the Constructed Environment and was vice chairman for the U.S. National Committee on Tunneling Technology. Other positions held were president, U.S. Committee on Large Dams; chair, Committee on Dam Safety, International Commission on Large Dams; executive committee, Construction Industry Institute; and the board of directors, Research and Management Foundation of the American Consulting Engineers Council. He has numerous professional affiliations, including fellowships in the American Society of Civil Engineers and in the Society of American Military Engineers. He holds a B.S. degree in civil engineering from the University of Minnesota, which awarded him the Board of Regents Outstanding Achievement Award.

**G. Brian Estes** is the former director of construction projects at Westinghouse Hanford Company, where he directed project management functions supporting operations and environmental cleanup of the Department of Energy Hanford nuclear complex. He was appointed as a member of this committee because of his experience with DOE, as well as other large-scale government construction and environmental restoration projects. He served on the committee that produced the recent NRC report *Improving Project Management in the Department of Energy* and has served on a number of other NRC committees. Prior to joining Westinghouse, he completed 30 years in the Navy Civil Engineer Corps, achieving the rank of rear admiral. Admiral Estes served as commander of the Pacific Division of the Naval Facilities Engineering Command and as commander of the Third Naval Construction Brigade at Pearl Harbor. He supervised over 700 engineers, 8,000 Seabees, and 4,000 other employees in providing public works management, environmental support, family housing support, and facility planning, design, and construction services. As vice commander, Naval Facilities Engineering Command, Admiral Estes led the total quality management transformation at headquarters and two updates of the corporate strategic plan. He directed execution of the $2 billion military construction program and the $3 billion facilities management program while serving as deputy commander for facilities acquisition and deputy commander for public works, Naval Facilities Engineering Command. He holds a B.S. in civil engineering from the University of Maine, an M.S. in civil engineering from the University of Illinois, and is a registered professional engineer in Illinois and Virginia.

**David N. Ford** is an assistant professor of civil engineering at Texas A&M University. He serves as a member of this committee because of his expertise in evaluating project management with analytical methods and simulations. He researches the dynamics of project management and the strategy of construction organizations, as well as teaching project management and computer simulation courses. Current research projects include an investigation into the causes of failures to implement fast-track processes and the value of contingent decisions in project strategies. Prior to his appointment at Texas A&M, Dr. Ford was an associate professor in the Department of Information Sciences at the University of Bergen in Norway. He was one of two professors to develop and lead the graduate program in the system dynamics methodology for 4 years. Dr. Ford's research during this time focused on the dynamics of product development processes and included work with Ericsson Microwave to improve that company's product development processes. Dr. Ford designed and managed the development and construction of facilities during 14 years in professional practice for owners, design professionals, and builders. The projects varied in size and facility type, including commercial buildings, residential development, industrial, commercial, and defense facilities. He serves as a reviewer for the journals *Management Science, Journal of the Operational Research Society, Technology Studies,* and *System Dynamics Review.* Dr. Ford received his B.C.E. and M.E. degrees from Tulane University and his Ph.D. from the Massachusetts Institute of Technology in dynamic engineering systems.

**G. Edward Gibson, Jr.,** is a professor of civil engineering, associate chairman for architectural engineering, and the Austin Industries Endowed Faculty Fellow in the Construction Engineering and Project Management program at the University of Texas at Austin. He serves as a member of this committee because of his expertise and research in preproject planning, organizational change, and the development of continuing education training programs for project managers. His research interests include organizational change, preproject planning, construction productivity, international project risk management, electronic data management, and automation and robotics. Dr. Gibson is a codirector of the Center for Construction Industry Studies funded by the Alfred P. Sloan Foundation. He received the Outstanding Researcher Award of the Construction Industry Institute (CII) for his pioneering work in preproject planning and is an author or coauthor of numerous articles and reports on this subject, including the CII *Pre-Project Planning Handbook* and the CII *Project Definition Rating Index* (PDRI). He also developed several CII education modules for continuing education and has taught over 140 short courses to industry in such areas as objective setting, team alignment, continuous improvement, preproject planning, and materials management. He received an M.B.A. from the University of Dallas and a B.C.E. and a Ph.D. in civil engineering from Auburn University.

**Theodore C. Kennedy** (National Academy of Engineering) is chairman and cofounder of BE&K, a privately held international design-build firm that provides engineering, construction, and maintenance for process-oriented industries and commercial real estate projects. Mr. Kennedy serves as a member of the committee because of his experience and expertise with the design, construction, and cost estimation of complex construction and engineering projects. BE&K

companies design and build for a variety of industries, including pulp and paper, chemical, oil and gas, steel, power, pharmaceuticals, and food processing. BE&K is consistently listed as one of *Fortune* magazine's Top 100 Companies to Work For, and BE&K and its subsidiaries have won numerous awards for excellence, innovation, and programs that support its workers and communities. Mr. Kennedy is the chairman of the national board of directors of INROADS, Inc., and is a member of numerous other boards, including the A+ Education Foundation and the Community Foundation of Greater Birmingham. He is also a member of the Duke University School of Engineering Dean's Council and the former chairman of the Board of Visitors for the Duke University School of Engineering. He is the former president of Associated Builders & Contractors and the former chairman of the Construction Industry Institute. He has received numerous awards, including the Distinguished Alumnus Award from Duke University, the Walter A. Nashert Constructor Award, the President's Award from the National Association of Women in Construction, and the Contractor of the Year award from Associated Builders and Contractors. Mr. Kennedy has a B.S. in civil engineering from Duke University.

**Michael A. Price** is manager of education programs for the Project Management Institute (PMI), an international association of project management professionals that provides accreditation and training. He was appointed to this committee because of his experience and expertise in developing and evaluating project management training programs. Dr. Price is responsible for the development and implementation of operational plans for all PMI educational programs and initiatives, including accreditation of degrees in project management; selection and coordination of 150 public seminars annually; management of continuing education requirements and record keeping for 22,000 project management professionals; and identification of new educational products and programs to meet the learning needs of the global project management community. Previous to his present position, Dr. Price was director of professional practice for the American Institute of Architects (AIA) and director of programs for architecture and engineering with the Research Center for Continuing Professional and Higher Education at the University of Oklahoma. He is an active member of the AIA and has been a member of the Education System Audit Review Task Group and the site visitation team for the National Architectural Accreditation Board. Dr. Price has a B.S. in environmental design, a B.Arch., an M.Ed., and a Ph.D. from the University of Oklahoma.

# Committee Fact-Finding and Briefing Activities and Documents Reviewed, February 2003 Through June 2003

## FACT FINDING AND BRIEFINGS

February 5-7  Committee meeting 12, Savannah River Site, Aiken, South Carolina. Roundtable discussion of current project management issues (implementation of O 413.3, project definition for Office of Environmental Management (EM) activities, professional development and core competencies, front-end project planning, risk analysis and management, earned value management systems (EVMSs) and tracking project data, performance-based contracting, etc.). Sessions will focus on issues raised by the following Savannah River Operations (SRO) and Westinghouse Savannah River Company (WSRC) managers:

> John Phillips, Project Management Program Manager, SRO
> Charlie Hansen, Deputy Manager, SRO
> Jeff Allison, Manager, SRO
> Ed Wilmot, Manager, Savannah River Defense Programs
> Bruce Wilson, Manager, Savannah River Defense Programs
> Sterling Franks, Manager, Savannah River Nuclear Nonproliferation Projects
> Bob Pedde, President WSRC
> Bill Elkins, Vice President for Projects, Design, and Construction, WSRC

Overview of WSRC Disciplined Conduct of Project Improvement
Initiatives
Jon Lunn, Manager, WSRC

Briefings on Savannah River Site (SRS) projects

Tritium Consolidation and Modernization Project
Pete Kozak, Federal Project Manager, SRO
Cleaning and Loading Line Modifications Project
Wayne Leslie, Federal Project Manager, SRO
Overview of Tritium Extraction Facility (TEF) Project
Clay Ramsey, Director, TEF Project Office, SRO
Mike Hickman, Federal Project Manager, SRO
Salt Waste Processing Facility Project
Terry Spears, Federal Project Manager, SRO
Chiller Replacement Project
Eddie Hipp, Federal Project Manager, SRO
Glass Waste Storage Building Project
Kim Sidey, Federal Project Manager, SRO
K Area Material Storage Project
Robert Barnes, Federal Project Manager, SRO
Pu Packaging and Stabilization Project
Guy Girard, Federal Project Manager, SRO

Roundtable discussion of project management manual and related
project management issues

Willie Clark, National Nuclear Security Administration (NNSA),
Office of Project Management and Systems Support (NA-54)
Roland Frenck, NNSA, NA-54

April 7        Informal meetings with senior managers

James Campbell, Director, Office of Management and Budget
Evaluation (OMBE)
James Powers, Director, Office of Program Analysis and Evalu-
ation (PA&E)
James Rispoli, Director, Office of Engineering and Construc-
tion Management (OECM)
Robert Card, Undersecretary

April 8-10    Committee meeting 13, Keck Center, Washington, D.C. Updates on
              project management process improvement and project performance

              Office of Engineering and Construction Management (OECM)
              James Rispoli, Director
              Robert McMullan, Deputy Director
              Mike Donnelly, Engineer
              David Treacy, Engineer
              Thad Knopnicky, Engineer
              Office of Science (SC), Construction Management Support
              Division (SC-81)
              Daniel Lehman, Director
              James Carney, Deputy Director
              NNSA, NA-54
              Willie Clark, Director
              Roland Frenck, Deputy Director
              EM, Office of Project Management (EM-6)
              Jay Rhoderick, Director

              Roundtable discussion of current project management issues
              (Manual M 413.3-1, definition and application of tailoring O 413.3
              requirements; roles and responsibilities of headquarters and field
              managers for critical decisions and project reviews; DOE project
              management culture: change and permanence, process and expecta-
              tions of risk management, project management education and career
              development programs; Project Analysis and Reporting System
              [PARS] data quality)

              Representatives of OMBE, OECM, EM, NNSA, and SC

June 4-6      Informal briefing and discussions, Albuquerque, New Mexico,
              with NNSA, Sandia Site Office (SSO), Los Alamos Site Office
              (LASO), and Energy Facility Contractors Group (EFCOG)

              *Project Management for the Acquisition of Capital Assets*
              (M 413.3-1) March 28, 2003, rollout
              Willie Clark, NNSA, Office of Project Management and
              Systems Support
              Roland Frenck, NNSA, Office of Project Management and
              Systems Support

Sandia Site Office (SSO)
Roles and Responsibilities of Federal Project Directors
Strategic Planning/Preproject Planning
Budgeting and Site Project Planning
Training and Career Development, Project Management Career
Development Program (PMCDP) Initiatives, SSO Quali-
fication Process
SSO Staffing
General Discussion, Planning, AEP, IPTs, Risk Manage-
ment, Documentation
Mike McFadden, Assistant Manager, Facilities and Infra-
structure, SSO
Frank White, Assistant Manager, Facilities and Infra-
structure, SSO

Sandia National Laboratories (SNL) Projects
Joint Computational Engineering Laboratory
Tom Goss, Project Director, SSO
Test Capabilities Revitalization
Wayne Evelo, Project Director, SSO
Center for Integrated Nanotechnologies
Frank White, Assistant Manager, Facilities and Infra-
structure, SSO

Los Alamos Site Office (LASO)
Roles and Responsibilities of Federal Project Directors
Training and Career Development, PMCDP Initiatives,
Qualification Process
Budgeting and Site Project Planning
General Discussion, Planning, AEP, IPTs, Risk Manage-
ment, Documentation
Herman LeDoux, Assistant Manager for Project Manage-
ment, LASO
Strategic Planning/Preproject Planning
Steve Fong, Federal Project Director, LASO

Los Alamos National Laboratory (LANL) Projects
Cerro Grande Fire
Jose Cedillos, Federal Project Director, LASO
Everett Trollinger, Federal Project Director, LASO
Chemistry and Metallurgy Research Facility Replacement
Juan Griego, Federal Project Director, LASO
Nonproliferation and International Security Center
Steve Fong, Federal Project Director, LASO

Energy Facility Contractors Group (EFCOG) discussions
SNL Construction Overview
Dave Corbett, Manager, SNL
MESA Project and the Effects of Manual M 413.3-1
Don Cook, Manager, SNL
Jennifer Girand, Manager, SNL
Test Capability Revitalization Project and the Effects of M 413.3-1
Lynne Schluter, Manager, SNL

EFCOG Desired Changes to M 413.3-1
Frank Figueroa, Vice President, SNL

Effect of 413.3 Manual on Other Ongoing Projects at Selected Sites
Lynne Schluter, Sandia National Laboratories (SNL)
David Chastain, Los Alamos National Laboratory (LANL)
John Shaffer, National Renewable Energy Laboratory (NREL)
Sam Formby, Savannah River Site (SRS)
Tom Etheridge, Oak Ridge National Laboratory (ORNL)

June 11-13    Informal meetings, Washington, D.C.

Office of Science
Raymond Orbach, Director
James Decker, Principal Deputy Director
Daniel Lehman, Construction Management Support Division
James Carney, Construction Management Support Division
National Nuclear Security Administration
Linton Brooks, Administrator
Greg Rudy, Associate Administrator, Facilities and Operations
Willie Clark, Office of Project Management and Systems Support
Roland Frenck, Office of Project Management and Systems Support
Environmental Management
Jessie Roberson, Assistant Secretary, EM
Paul Golan, Chief Operating Officer, EM
Jay Rhoderick, Director, Office of Project Management

# DOCUMENTS REVIEWED

## Savannah River Site

DOE-SRS Project Management Manual (SRM 410.1.1B), April 21, 2001
Disciplined Conduct of Projects, Causal Analysis and Corrective Action Plan
(G-ESR-G-00046 Rev 0), March 31, 2002
Risk Management Plan for the Plutonium Packaging and Stabilization Project
(Y-RMP-F-00004 Rev 0), August 21, 2000
Cleaning and Loading Modifications (CALM) Project M03A/B, Conceptual
Design Risk/Opportunity Assessment Report CDR2, January 31, 2003

## Testimony Before the U.S. House of Representatives Committee on Government Reform, March 20, 2003

Breathing Fumes: A Decade of Failure in Energy Department Acquisitions,
Tom Davis, Chairman, U.S. House of Representatives Committee on
Government Reform
Statement of Gregory H. Friedman, Inspector General, DOE
Statement of James Rispoli, Director, OECM
Status of Contract and Project Management Reforms, Statement of Robin M.
Nazzaro, General Accounting Office

## Related Reports

GAO. Major Management Challenges and Program Risks, Department of
Energy, GAO-03-100, January 2003
DOE IG. Management Challenges at the Department of Energy, December 2002
DOE IG. Progress of the Spallation Neutron Source Project, November 2001
DOE IG. Dual Axis Radiographic Hydrodynamic Test Facility, May 2003
Project Management Institute, Organizational Project Management Maturity
Model (OPM3), 2003

## Spreadsheets

List of current DOE capital acquisition projects
List of baseline changes
Summary External Independent Review of Critical Decision-2 Baseline
Validation FY2001 Through FY2003

## Project Planning Documents

Los Alamos National Laboratory, Chemistry and Metallurgy Research Facility Replacement Project, Mission Need Statement, Revision 0, February 25, 2002

Elimination of Weapons-Grade Plutonium Production Program, Justification of Mission Need for Seversk Plutonium Production Elimination Project, Zheleznogorsk Plutonium Production Elimination Project, and Nuclear Safety Upgrades Project, December 16, 2002

Columbia River Corridor Project, Justification of Mission Need, Revision 0, July 2002

Columbia River Corridor, Project Execution Plan, Revision 0, November 2002

National Energy Technology Laboratory (NETL) Technology Support Facilities, Project Plan (Revised Critical Decision 0), November 2002

**External Independent Reviews (EIRs) and Independent Cost Reviews (ICRs)**

**Integrated Closure Project Baseline**

Rocky Flats, Burns and Roe (B&R), June 2001

**EIRs/ICRs CD-2 Reviews**

Argonne National Laboratory (ANL), Mechanical Systems Upgrade, Jupiter, September 2002

Idaho Engineering and Environmental Laboratory (INEEL), INTEC Cathodic Protection and Expansion, Logistics Management Institute (LMI), August 2001

INEEL, Glove Box Evacuator, Jupiter, July 2002

INEEL, CERCLA Disposal Facility, Jupiter, March 2002

Kansas City Plant, Gas Transfer Capacity Expansion, 03-D-121, LMI, November 2002

Lawrence Livermore National Laboratory (LLNL), Engineering Technology Center Upgrades, 02-D-105, LMI, July 2002

LLNL, Sensitive Compartmented Information Facility, Jupiter, July 2002

Los Alamos National Laboratory (LANL) SM-43 Replacement, LMI, December 2001

LANL, TA-50 Waste Management Risk Mitigation, Jupiter, September 2002

Mound Site, Potential Release Site, B&R, December 2002

Nevada Test Site (NTS), Bus Upgrades Project, LMI, September 2002

NTS, Atlas Relocation, Jupiter, August 2001

Oak Ridge National Laboratory (ORNL), Research Support Center, MEL-001-25, LMI, August 2002

ORNL, Center for Nanophase Sciences, Jupiter, August 2002
Pacific Northwest National Laboratory (PNNL), Laboratory Systems Upgrades,
    D-494 (MEL-001-18), LMI, May 2002
Pantex, Weapons Evaluation Test Laboratory, 01-D-126, LMI, May 2002
Sandia National Laboratories (SNL), Test Capabilities Revitalization Project
    (Phase I), LMI, December 2002
SNL, Joint Computational Engineering Laboratory (JCEL), 00-D-107, LMI,
    September 2001
SNL, Underground Reactor Facility (SURF), Jupiter, October 2002
SNL, Microsystems and Engineering Sciences Applications (MESA), 01-D-
    108, LMI, November 2001, B&R, February 2003
SNL, Mixed Waste Landfill, Jupiter, July 2001
SNL/Livermore, California, Distributed Information Systems Laboratory
    (DISL), DP 01-D-101, LMI, July 2001
Savannah River Site (SRS), Western Sector Dynamic Underground Stripping
    (DUS) Project, LMI, October 2002
SRS, Pit Disassembly and Conversion Facility, B&R, May 2002, October 2002
SRS, Plutonium Packaging and Stabilization, Jupiter, April 2002
SRS, Highly Enriched Uranium Blend Down, Jupiter, November 2001
SRS, Mixed Oxide Fuel Fabrication Facility, Jupiter, February 2002
Stanford Linear Accelerator Center (SLAC), Gamma-ray Large Area Space
    Telescope
Large Area Telescope (GLAST), Jupiter, August 2002
Y-12, Special Materials Purification Facility, Jupiter, September 2002


**EIRs/ICRs CD-3 Reviews**

Hanford Site, Washington, Waste Treatment and Immobilization Plant Project,
    LMI, September 2002
Los Alamos National Laboratory, Replacement Emergency Operations Center,
    Jupiter, July 2001
Lawrence Livermore National Laboratory, Terascale Simulation Facility, 00-D-
    103, LMI, August 2001


**Pre-O 413.3 Reviews**

Brookhaven National Laboratory, Groundwater and Surface Water Protection
    Upgrades, SC 01-CH-103-0, LMI, August 2000
Hanford Site Office of River Protection, Initial Tank Retrieval System, Jupiter,
    May 2000
Hanford Site, Immobilized High Level Waste Storage, Jupiter, May 2000
Idaho National Engineering and Environmental Laboratory (INEEL), Health
    Physics Instrumentation Laboratory, LMI, June 1999

INEEL, Sitewide Information Network, EM 01-D-404, LMI, April 2000

INEEL, Advanced Mixed Waste Treatment, LMI, April 2000

INEEL, Site Operations Center, LMI, April 2000

INEEL, INTEC Cathodic Protection System Expansion Project, EM 01-ID-402, LMI, August 2000

INEEL, Test Reactor Area Electrical Upgrade, June 1999

Lawrence Livermore National Laboratory (LLNL), Terascale Simulation Facility, LMI, July 1999

Los Alamos National Laboratory (LANL), Strategic Computing Complex, LMI, July 1999

LANL, Nonproliferation and International Security Center (NISC), LMI, 1999

LANL, Isotope Production Facility, LMI, August 1999

Oak Ridge National Laboratory (ORNL), Laboratory Facilities HVAC Upgrade, SC MEL-001-15, LMI, August 2000

ORNL, Laboratory for Comparative and Functional Genomics, SC 01-E-300, LMI, August 2000

ORNL, Spallation Neutron Source, B&R, December 1999

Sandia National Laboratories (SNL)/California, Distributed Information Systems Laboratory, DP 01-D-101, LMI, June 2000

SNL, Joint Computational Engineering Laboratory, LMI, May 2000

SNL, Microsystems and Engineering Sciences Applications (MESA), 01-D-108, LMI, August 2000

Stanford Linear Accelerator Center (SLAC), Research Building, LMI, June 1999

Waste Isolation Pilot Plant (WIPP), Remote Handled Transuranic Waste Transportation, Jupiter, May 2000

Y-12, Highly Enriched Uranium Materials Facility, Jupiter, June 2000

## Other Reviews, Independent Cost Estimates (ICEs), and Independent Cost Reviews (ICRs)

Portsmouth Gaseous Diffusion Plant, Emergency Winterization/Cold Lay-Up Project, B&R, February 2001

Savannah River Site (SRS), Tritium Facility Modernization and Consolidation, B&R, September 2000

Civilian Radioactive Waste Management (CRWM) Program Capital Asset Plan, B&R, 2002

CRWM Program, 2001 Total System Life Cycle Cost (TSLCC), B&R, January 2001

Yucca Mountain Project, License Application to Initial Operating Capability, B&R, January 2003 and August 2002

West Jefferson Site, Independent Cost Review, Columbus Environmental Management Project (CEMP) Baseline, B&R, November 2001

Fernald Site Environmental Management Project (FEMP) Closure Contract
Baseline (CCB), ICR, B&R, August 2001
Lawrence Livermore National Laboratory, National Ignition Facility Project,
Revised Cost and Schedule Baselines, B&R, September 2000
Fermi National Accelerator Laboratory, Neutrinos at the Main Injector (NuMI),
Project Proposed Baseline, B&R, November 2001
West Valley Demonstration Project, Proposed Path to Completion Plan, B&R,
November 2001

# APPENDIX D

# Compilation of Findings and Recommendations

The three previous NRC reports (NRC, 1999, 2001, 2003) include 76 findings and 94 recommendations that are compiled below. The previous findings and recommendations as well as those in this report address 10 recurring objectives for the changes needed to improve project management at DOE. Most of these changes relate to inadequate planning, inadequate risk management, and inadequate monitoring and follow-up. The recurring objectives include the following:

- Develop policies and procedures to define the DOE method of managing projects;
- Create a project management culture across the agency that supports the consistent implementation of policies and procedures;
- Provide leadership that ensures disciplined planning and execution of projects as well as support for continuous process improvement;
- Provide a project management champion at the highest level of the department to ensure that a focus on the importance of project management is established and maintained;
- Develop competence in fulfilling the owner's role in front-end project planning, risk management, and project execution;
- Apply rigorous project reporting and controls that include earned value systems, link day-to-day management data to periodic reporting, forecast time and cost to complete, and maintain historical data with which to benchmark project performance;
- Document processes and performance to support benchmarking and trend analysis;

- Invest in human capital by providing training and career development to ensure an adequate supply of qualified, skilled project directors;
- Continue, refine, and document a program of external and internal project reviews; and
- Employ innovative approaches to capital acquisition and the use of performance-based contracting.

Although the committee has provided suggested approaches, it believes that DOE senior managers have the responsibility to identify and apply metrics to define the appropriate level of excellence and to drive continuous process improvement.

## POLICIES, PROCEDURES, DOCUMENTATION, AND REPORTING

### 1999

**Finding.** DOE does not have adequate policies and procedures for managing projects. No single authority is responsible for enforcing or ensuring that project management tools are used.

**Finding.** DOE has developed comprehensive practice guidelines for the design and construction phases of projects but has not developed comparable guidelines for the early conceptual and preconceptual phases, when the potential for substantial savings is high.

**Finding.** Many DOE projects do not have comprehensive project management plans to define project organization, lines of authority, and the responsibilities of all parties.

**Finding.** DOE does not effectively use value engineering to achieve project savings, even though federal agencies are required to do so.

**Finding.** DOE project documentation is not up to the standards of the private sector and other government agencies.

**Finding.** DOE does not have a consistent system for controlling changes in project baselines.

**Finding.** DOE does not effectively use available tools, such as earned value management, to track the progress of projects with respect to budget and schedule.

**Finding.** ISO 9000 provides a certification process by which an organization can measure itself against its stated goals, but DOE has not obtained certification.

The certification process would help DOE remake the entrenched operating procedures and standards that have accumulated over the past 50 years.

**Recommendation.** As a part of its project management system, DOE should issue fundamental policies, procedures, models, tools, techniques, and standards; train project staff in their use; and require their use on DOE projects. DOE should develop and support the use of a comprehensive project management system that includes a requirement for a comprehensive project management plan document with a standard format that includes a statement of the project organization covering all participating parties and a description of the specific roles and responsibilities of each party.

**Recommendation.** DOE should update the project performance studies to document progress in these areas and extend the benchmarking baseline to include all major DOE construction projects. The study results should then be used to improve project procurement and management practices.

**Recommendation.** DOE should mandate a reporting system that provides the necessary data for each level of management to track and communicate the cost, schedule, and scope of a project.

**Recommendation.** DOE should establish a system for managing change that provides traceability and visibility for all baseline changes. Change control requirements should apply to the contractor, the field elements, and headquarters.

**Recommendation.** DOE should establish minimum requirements for a cost-effective earned-value performance measurement system that integrates information on the work scope (technical baseline), cost, and schedule of each project. These requirements should be included in the request for proposals.

**Recommendation.** DOE, as an organization, should obtain and maintain ISO 9000 certification for all of its project management activities. To accomplish this, DOE should name one office and one individual to be responsible for acquiring and maintaining ISO 9000 certification for the whole department and should require that consultants and contractors involved in the engineering, design, and construction of projects also be ISO 9000 certified.

**Recommendation.** DOE should establish an organization-wide value-engineering program to analyze the functions of systems, equipment, facilities, services, and supplies for determining and maintaining essential functions at the lowest life-cycle cost consistent with required levels of performance, reliability, availability, quality, and safety. Value engineering should be done early in most projects, and project managers should take the resulting recommendations under serious consideration.

## 2001

**Finding.** The recommendations in the Phase II report to develop and publish a set of policies and procedures for management of DOE projects appear to have been addressed to some extent by the draft [Program and Project Management manual] PPM and the draft [Project Management Practices] PMP; however, the committee finds that there is a need for additional detail and clarity and elimination of discontinuities, gaps, overlaps, and repetitions. The committee recognizes that OECM is addressing these issues as it develops the next iterations and commends OECM for its leadership role.

**Recommendation.** The PPM and PMP text should be tailored to specific DOE requirements. It should be clear which parts of the text constitute DOE required procedures and which parts reflect general advice on good project management practices.

**Recommendation.** OECM should assure that policies and required procedures add value by streamlining the process and improving project performance. Policies and procedures that do not demonstrably add value should be revised or eliminated.

**Recommendation.** The PPM and PMP should have parallel structures. A complete index and a glossary of terms should be provided for both documents.

**Recommendation.** Examples should be given where they will illustrate the application of procedures and the necessary documentation. Examples should have adequate explanations and represent realistic project situations. Over time, a set of templates and case studies should be built up.

**Recommendation.** OECM should be provided the resources needed to publish improved, revised versions of the PMP and PPM as soon as possible. OECM should be given the authority to authorize case-by-case exceptions when appropriate to ensure that common sense and cost-effectiveness prevail in the retrofitting of procedures to ongoing projects.

## 2002

**Finding:** DOE and DOE contractor personnel expressed some concern that requirements in DOE Order O 413.3 and drafts of the *Program and Project Management* (PPM) manual and *Project Management Practices* (PMP) result in excessive and unnecessary effort and cost for projects of less than $20 million; they believed these requirements should not apply to environmental projects and that front-end planning documentation and review requirements are excessive. The committee does not agree with their views but it does believe that requirements should be tailored to the complexity of the project.

**Recommendation:** DOE should resist efforts that reduce requirements for front-end planning and the critical decision-review process. This resistance is necessary to ensure that the process is uniform and that projects selected for execution are consistent with DOE's strategic plan. The requirements should apply to all projects over $5 million and be tailored to the complexity of the project.

**Finding:** The committee has observed examples of both effective and ineffective project management practices at DOE. Order O 413.3 is intended to create a consistent department-wide definition of what is required of DOE project managers. OECM is revising the PPM to better define the project management practices to be used to achieve the objectives of O 413.3. The committee believes that the order is beginning to increase the level of consistency throughout the department. It also believes that a document to define the minimum actions required to implement O 413.3 is necessary. This document needs to carry commensurate authority and be coordinated to develop a consistent DOE approach to project management and project oversight.

**Recommendation:** OECM should accelerate development of the PPM and should issue the current draft immediately to guide interim compliance with O 413.3. The order and the manual should be separate but coordinated documents to create a consistent DOE approach to project management. The order should continue to define what is required and remain relatively unchanged over time. The manual should continue to be a separate document to specify minimum requirements for compliance to O 413.3.

## PROJECT PLANNING AND CONTROLS

### 1999

**Finding.** DOE preconstruction planning is inadequate and ineffective, even though preconstruction planning is one of the most important factors in achieving project success.

**Finding.** DOE often sets project baselines too early, usually at the 2- to 3-percent design stage, sometimes even lower. (An agreement between Congress and DOE's chief financial officer for establishing baselines at the 20- to 30-percent design stage is scheduled to be implemented in fiscal year 2001.)

**Finding.** DOE often sets project contingencies too low because they are often based on the total estimated cost of a project rather than on the risk of performing the project.

**Finding.** DOE does not always use proven techniques for assessing risks of major projects in terms of costs, schedules, and scopes.

**Recommendation.** DOE should require that strategic plans, integrated project plans, integrated regulatory plans, and detailed project execution plans be completed prior to the establishment of project baselines. To ensure facility user and program involvement in the preconstruction planning process, DOE should require written commitments to project requirements from the ultimate users.

**Recommendation.** DOE should significantly increase the percentage of design completed prior to establishing baselines. Depending on the complexity of the project, the point at which project baselines are established should be between the completion of conceptual design and the completion of the preliminary design, which should fall between 10 and 30 percent of total design. The committee supports continuing efforts by Congress and the DOE to develop project baselines at a point of adequate definition beginning with fiscal year 2001.

**Recommendation.** Baseline validation should be assigned specifically to the project management office recommended in this report. The Military Construction Program of the U.S. Department of Defense, which requests planning and design funds for all projects in the preliminary design stage on the basis of total program size, is a potential model for DOE.

**Recommendation.** DOE should establish contingency levels for each project based on acceptable risk, degree of uncertainty, and confidence levels for meeting baseline requirements. The authority and responsibility for managing contingencies should be assigned to the project manager responsible for doing the work. In the process of evaluating potential projects, DOE should apply risk assessment and probabilistic estimating techniques, as required by the Office of Management and Budget.

## STRATEGIC PLANNING

### 2001

**Finding.** There are no PSO strategic plans defining long-range goals and objectives or mission needs, and documentation of project justification is almost entirely lacking or inadequate, so that it is impossible to assess whether the right projects are being done. Cost-benefit analysis and performance measurements required by GPRA cannot be performed effectively without effective strategic plans.

**Recommendation.** The PSOs should develop budget-based rolling 5-year strategic plans that identify the mission goals and objectives of the program, the projects necessary to achieve them, and the benefits to be expected from these projects.

## FRONT-END PLANNING

### 2001

**Finding.** The DOE process for project initiation, planning, justification, and execution continues to need substantial improvement. A top-to-bottom process that recognizes best practices in both government and industry, as well as the unique and specific requirements of DOE programs and projects, is essential.

**Recommendation.** The DOE secretarial acquisition executive should sponsor a process improvement program, and OECM should be named the program champion in DOE.

**Finding.** Compliance with the front-end planning requirements in O 413.3 has been inconsistent among PSOs and among individual projects.

**Recommendation.** OECM should assure that all program offices have a documented front-end planning process that meets the intent of O 413.3, and that the information used as input for Energy Systems Acquisition Advisory Boards (ESAABs) and ESAAB-equivalent readiness reviews, as well as the outcomes of these reviews, is documented and used to assess project performance and progress in improving project planning.

**Recommendation.** The PMSOs should consider developing tailored checklists such as the EM-PDRI [Office of Environmental Management Project Definition Rating Index] as in-process planning tools, train project personnel to use them, and analyze their effectiveness for projects throughout the DOE complex. Effective and consistent front-end planning should be made mandatory for all projects.

**Finding.** Tools such as checklists, communications software/methods, planning reviews, third-party audits, economic modeling, objective setting, and team building, if used correctly, can contribute to effective front-end planning. Performance of technical evaluation during planning is essential for projects involving new technology, complex site conditions, and complex project-flow requirements. Consistent documentation and planning structure would increase the effectiveness of front-end planning in the department.

**Recommendation.** OECM should clarify, expand, and revise the front-end planning procedures in the *Program and Project Management* manual and *Project Management Practices*. DOE should use standard industry procedures where applicable; however, the PMSOs should provide supporting policies and procedures tailored to the specific projects and needs of each program. The PMSOs and OECM should assure the adequacy of front-end project planning prior to each critical decision, to assure that projects are not unnecessarily delayed by poor plans and that time constraints do not cause projects to be approved without adequate planning.

**Recommendation.** The deputy secretary and the designated program acquisition executives should strengthen their interest and support, thereby confirming that truly effective front-end planning will be required without exception. OECM and the PMSOs should pay close attention to documentation of front-end planning decisions.

**Finding.** DOE has established a process to significantly increase the accuracy and reliability of project baselines.

**Recommendation.** OECM should actively participate in the process and monitor the performance of projects baselined under this new process to document its impact and opportunities for improvement.

**Finding.** Overall, insufficient attention from DOE management is being given to the front-end planning process; however, the committee observed that management was acting in isolated cases and to varying degrees within the program offices.

**Recommendation.** DOE senior management should emphasize the importance of thorough and complete front-end planning (including written documentation). ESAABs and ESAAB-equivalent reviews should be used to enhance the quality of front-end project planning and assure that the project team is pursuing the right project—that is, that the project has adequate justification and will satisfy a well-conceived need.

**Finding.** Front-end planning improvement requires metrics for trend analysis. The committee was not able to obtain this information for specific projects because DOE does not have enough data for front-end planning trend analysis.

**Recommendation.** OECM should begin benchmarking project practices and performance metrics to identify areas in need of improvement and establish a baseline for future evaluation. This benchmarking effort should be systematic, quantitative, and analytical, and it should compare practices in industry and in other government agencies. It should capture both front-end planning and performance metrics, including actual performance versus forecast.

**Finding.** A training program addressing front-end planning and other project management practices is being developed. The completion date of this effort was reported to the committee to be December 2002, with training to start soon afterward. Without immediate improvement in the planning knowledge and skills of personnel and more management emphasis on improving the planning process, projects will continue to have inadequate front-end planning.

**Recommendation.** OECM should do more than develop policies and procedures—it should become fully engaged in process improvement beginning with front-end planning. To overcome the lack within the department of skilled project planners and the delays in training, and to bridge the gap until a training program takes effect, DOE should establish a cadre of experienced project planners within OECM; they should have a wide variety of planning capabilities and prior experience in different project types, including high-risk projects. These individuals should be a part of the initial integrated project teams and should assist the project originators (as internal consultants) in getting front-end planning done correctly, including planning prior to CD-0. This cadre of internal consultants should champion the DOE front-end planning process, providing just-in-time training for front-end planning to project teams. DOE should benchmark its management of project planning personnel and application of their expertise with that of private sector companies that have successfully undertaken similar activities. In this way, DOE may be able to jump-start an immediate improvement in planning capability.

**Recommendation.** DOE should eliminate impediments to initiating training for front-end project planning prior to December 2002. Training should begin as soon as possible.

## RISK MANAGEMENT

### 2001

**Finding.** With rare exceptions, there are no risk models for ongoing DOE projects, and back-fitting risk assessment to ongoing legacy projects does not seem to be part of the acquisition risk management (ARM) study. There is no consistent system for evaluating the relative risks of projects with respect to scope, cost, or duration, so the deputy secretary, the chief financial officer, and the PSO managers have no objective basis for knowing which projects are riskier (and therefore require more management attention) than others.

**Recommendation.** DOE should develop the ability to perform quantitative risk assessments. These assessments should be carried out by DOE personnel with experience in such analyses working with persons who have an in-depth understanding of a given project. Internal project risk assessments should be separately evaluated by independent assessors or reviewers who are not project proponents for reasonableness of assumptions, estimates, and results. Risk mitigation and management plans should be prepared that can deal with significant risks identified.

**Recommendation.** DOE project management personnel should be trained in risk assessment methodology. This training should cover not only risk analysis methodology and techniques, but also the managerial responsibilities related to interpretation of risk assessments and mitigation and management of risks.

**Recommendation.** Risk analyses should explicitly consider the interdependence of the various activities due to common modes (root causes), or document why there is no dependence.

**Finding.** DOE has not implemented statistical models (the "objective" analysis cited in OMB Circular No. A-94, *Guidelines and Discount Rates for Benefit-Cost Analysis of Federal Programs*) because it has no usable database of past and current projects.

**Recommendation.** DOE should develop an internal database of data on its own projects and on projects of other owners. A system should be established to capture data on current and future projects. Data on comparable projects performed by other federal agencies and by industry should be obtained and included. The current development of the Project Analysis and Reporting System (PARS) (discussed in Chapter 5) could be a step toward this goal, and the committee plans to follow this work with interest. Although its early stage of development prevents assessing its effectiveness at this time, the level of participation by projects, accuracy of data, completeness of data, and avoidance of duplication should be addressed by OECM. The architecture of this data system should be specifically designed to provide support for the analysis of risks for ongoing and future projects.

**Finding.** By and large, DOE's practices in risk assessment and risk management have not significantly improved since the Phase II report. The committee reviewed some project risk assessment studies but did not see an example of a risk assessment or risk mitigation plan that it finds acceptable. The discussion in the draft PPM is merely an outline, and the material in the draft PMP is not useful as a guide for practicing risk management. Conversely, the current ongoing acquisition risk management (ARM) pilot study at three DOE sites and by the Contract Reform and Privatization Office and the EM Division Steering Group/Working Group, due for completion by December 2001, is a positive move and shows promise. The committee intends to follow this study with interest as it evolves.

**Recommendation.** The current acquisition risk management (ARM) pilot study should be continued and expanded beyond budget risks to cover the issues addressed in the Phase II report and in this report, such as schedule, scope, quality, and performance risks.

**Finding.** DOE's deficiencies in risk analysis lead to inadequate risk mitigation planning and execution. Plans often address symptoms but not causes. Execution is typically reactive or nonexistent. To be useful during project implementation, this planning should, at a minimum, do the following:

- Characterize the root causes of major risks that were identified and quantified in earlier portions of the risk management process.
- Identify alternative mitigation strategies, methods, and tools for each major risk.
- Evaluate risk interaction effects.
- Identify and assign priorities to mitigation alternatives.
- Select and commit required resources to specific risk mitigation alternatives.
- Communicate planning results to all project participants for implementation.

**Recommendation.** DOE should develop and implement risk mitigation planning processes and standards. Project risk assessment and management should be carried out throughout the project life cycle and should be part of the documentation for each critical decision point. Risk mitigation plans should be reviewed, critiqued, returned for additional work if needed, and approved by an independent organization such as the ESAABs at each critical decision point and prior to project approval for design or construction funding.

**Recommendation.** Until DOE project managers can be adequately trained in risk management, OECM should establish a cadre of experienced risk assessment personnel, who can be detailed or seconded to projects in the very early stages, to provide risk assessment expertise from the beginning of projects and incorporate risk management into the initial project management plan.

**Finding.** DOE needs to take a flexible approach in managing risk because of the high levels of uncertainty. To be effective in risk management, flexibility should be structured. A process is needed for designing, assessing, evaluating, and implementing risk-management alternatives that include decisions made during front-end project planning and decisions made after project initiation.

**Recommendation.** DOE should develop cutting-edge abilities to manage high-risk projects. It should adopt a process of identifying, designing, evaluating, and selecting risk management alternatives. The process should explicitly include and address alternatives that take advantage of opportunities for the partial resolution of important uncertainties after project initiation. Reviews at critical decision points should always entertain Plan B, that is, the alternative to be pursued if the primary approach is adversely affected by subsequent information or events.

**Finding.** An objective assessment is essential to performance-based contracting to assure that DOE does not shift to other project participants risks that it should retain or vice versa, or shift risks at more cost than they are worth.

**Recommendation.** DOE should explicitly identify all project risks to be allocated to the contractors and all those that it will retain, and these risks should be made known to prospective bidders. To use a market-based approach to allocating risks and to avoid unpleasant surprises and subsequent litigation, it is necessary that all parties to an agreement have full knowledge of the magnitude of risks and who is to bear them.

**Finding.** DOE project risks are not aggressively managed after project initiation. Risk management during projects is an inadequately developed project management capability at DOE.

**Recommendation.** DOE should initiate a program to improve the knowledge, skills, and abilities of project managers and develop tools and information needed to manage risk throughout the life of a project. Project participants who manage risks actively and achieve successful project performance should be appropriately rewarded.

**Finding.** The committee observed an ongoing deficiency in risk management that undermines DOE's ability to avoid surprises and take timely remedial action to avoid baseline breaches and to predict the actual cost to complete ongoing projects.

**Recommendation.** DOE should conduct an immediate and thorough risk assessment of all ongoing DOE projects with significant remaining time and costs. Such an assessment would establish, on a consistent basis, the risks and vulnerabilities of projects with respect to schedule, cost, and performance. It should assess the actual status of current projects and compare them with the project's original baselines, the current project schedules and budgets, and performance for comparable completed projects. The assessment should evaluate the risks of future scope shortfalls and budget and schedule overruns.

**Finding.** Innovative, cutting-edge, and exceptional risk management abilities are needed by DOE to identify and address the risks in many of its projects. DOE needs to develop expertise and excellence in managing very risky development projects. The DOE complex has the intellectual, computational, and other resources necessary to produce significant improvements in this area.

**Recommendation.** DOE should develop more expertise and improved tools for risk management. Nontraditional and innovative approaches, tools, and methods should be investigated for their adaptability to DOE project conditions and use in DOE risk management. They would include those cited earlier in this report and in the Phase II report (NRC, 1999, Appendix B), such as systems analysis, event trees, causal loop diagrams, system dynamics, and stochastic simulation, which have been tested and shown to be valuable on similar projects or in addressing similar challenges.

**Finding.** DOE does not seem to have a consistent or explicit policy on the use of management reserves, what size they should be, and who should control them.

**Recommendation.** The deputy secretary as secretarial acquisition executive, and the chief financial officer, assisted by the PSOs and OECM, should define and state DOE policy on management reserves. This policy should be clarified in a future release of O 413.3.

# FIRST-OF-A-KIND PROJECTS

## 2002

**Finding.** First-of-a-kind projects have been and can be successfully managed and executed by DOE, but they require particular care. The higher degree of uncertainty that attends these projects requires managers who are experienced in dealing with uncertainty and ambiguity. Not all project managers have this ability. The best project managers and management systems more than pay for themselves on first-of-a-kind projects by delivering projects on schedule with little budget overrun.

**Recommendation.** DOE managers and acquisition executives should pay particular attention to the unique characteristics of first-of-a-kind projects by considering the issues discussed above—for example, costs and benefits, scope, cost and schedule budgets, constructability, alternatives, management planning, and project controls—at all critical decision points.

## PROJECT CYCLE TIME

### 2002

**Finding.** Undersecretary Robert Card has enunciated a new strategy for Environmental Management (EM) that stresses earlier completion of site cleanup and remediation and earlier closure of sites or their turnover to private industry. The EM organization is reorganizing to fulfill this new strategy. Although it appears that much of the time reduction will be due to a reevaluation of the necessary end states, which may involve negotiations with stakeholders, the committee considers this initiative an important step toward DOE controlling its projects rather than being controlled by them, as has been the case. To make progress, it is necessary to believe that projects can be controlled and delivered earlier rather than believing that nothing can be done and that the process will require 70 years to complete. It is too early to determine whether the new EM organization will be successful, but the committee considers active attempts to get projects under control, to define strategic directions, and to align projects with strategy to be superior to passivity.

**Recommendation.** The strategy of achieving earlier completion of site remediation and closure or turnover of sites should, if successful, reduce environmental risks substantially and save U.S. taxpayers many billions of dollars. This initiative should be supported and continued.

**Recommendation.** In addition to redefining end states, DOE EM should consider all possible methods for improving its project management processes, preparing its project managers, and achieving earlier project completion, some of which are outlined above.

**Recommendation.** Program offices in DOE other than EM should also consider opportunities for earlier project delivery through application and implementation of the principles cited above.

## REPORTING AND OVERSIGHT

### 2001

**Finding.** The committee has observed some objections to changing or adding reporting practices to conform to a uniform system. Clearly, each organization in

DOE has become accustomed to its own reporting system, and these legacy systems should be taken into consideration when designing a new department-wide reporting system. Successful implementation of a management information system requires a knowledge and consideration of the needs and preferences of the users. A critical factor in the success of a reporting system is that those who are burdened with the input of data should also receive some benefits from doing so in an accurate and timely manner. For quality and consistency, it is necessary that each data element be input only once, as close to the source as possible. A schedule to phase in reporting requirements in a manner that does not disrupt ongoing projects or cause unnecessary costs may be needed.

**Recommendation.** DOE and its contractors should adopt full accrual cost accounting systems in order to provide EVMS (earned value management system) and PARS (Project Analysis and Reporting System) with appropriate data.

**Recommendation.** The PARS information system for collecting data from projects department-wide should be designed so that it supports the data needs of project managers to evaluate project performance as well as the oversight needs of the PSOs, the OECM, the CFO, and the deputy secretary. The database should also provide information for benchmarking future projects.

**Finding.** DOE management needs to be able to detect potentially adverse trends in project progress and distinguish them from mere random fluctuations in progress reporting. EVMS data provide some very valuable insights into the health of a project and can predict the probable outcome. They can also shed light on the conduct of the work, particularly when it is reported and analyzed to evaluate period-to-period trends.

**Recommendation.** DOE should utilize EVMS data to calculate the incremental and cumulative cost performance index (CPI), schedule performance index (SPI), and contingency utilization index (CUI) for each reporting period to analyze and improve project performance.

**Finding.** The committee had great difficulty in locating information documenting project decisions, the project management process, and project performance. [Defense Programs] DP is planning to participate in the CII benchmarking survey, but there is generally not enough consistent information to allow benchmarking project management performance within DOE or between DOE and other federal agencies and private industry.

**Recommendation.** All DOE projects should be benchmarked within DOE and between DOE and other federal agencies and private industry, and this benchmarking should be consistent across the department.

## SKILLS, SELECTION, AND TRAINING OF PERSONNEL

## 1999

**Finding.** DOE's failure to develop project management skills in its personnel is a fundamental cause of poor project performance. DOE has shown little commitment to developing project management skills, as indicated by the lack of training opportunities and the absence of a project management career path. Successful organizations recognize that project management skills are an essential core competency that requires continuous training.

**Recommendation.** DOE should establish a department-wide training program for project managers. To ensure that this program is realistic, practical, and state of the art, DOE should enlist the assistance of an engineer/construction organization with a successful record of training project managers. DOE should establish criteria and standards for selecting and assigning project managers, including documentation of training, and should require that all project managers be trained and certified. DOE should also require that all contractors' project managers be experienced, trained, and qualified in project management appropriate to the project.

## TRAINING AND PROFESSIONAL DEVELOPMENT

## 2001

**Finding.** Although there is a clear and immediate need to provide project management training, courses developed under the current [Project Management Career Development Program] PMCDP effort will not be available until late 2003. Training is the equivalent of providing workers the tools to accomplish their job.

**Recommendation.** DOE should immediately implement an accelerated training program to improve the knowledge, skills, and abilities of project managers to address recognized gaps while continuing the PMCDP planning effort. Immediate measures should be taken to eliminate impediments and use current resources to explore creative and cost-effective nonclassroom alternatives such as e-learning, action learning, and learning portfolios. Also, trainers skilled in specific topics should be engaged to instruct a cadre of DOE employees, who in turn will impart department-wide training to other DOE employees.

**Recommendation.** At the beginning of each fiscal year, DOE management should budget the funds to accomplish the projected training objectives for that year and should persist in mandating the accomplishment of individual career development objectives.

**Finding.** The existing contract for training offers a means to deliver consistent content throughout the department; however, it reduces the range of options for training.

**Recommendation.** DOE should modify or replace the current contract to allow greater flexibility in accessing courses pertinent to the project management skills utilized by industry and other federal agencies. DOE should develop new courses consistent with the new knowledge, skills, and abilities requirements identified by the findings of the gap analysis.

## HUMAN CAPITAL

### 2002

**Finding.** There is reason to believe, based on the reported numbers of DOE project management personnel and the volume of DOE projects, that DOE is understaffed in the area of project managers and essential project management support staff. The committee concludes that there may not be enough DOE project management personnel to discharge their responsibilities as the owner's representatives. This apparent deficiency may lead to a situation in which [management and organization] M&O and [management and integrations] M&I contractors, by design or default, are performing the roles and functions that should be the prerogative of the owner's representatives. This inappropriate devolution of some of the department's project management responsibilities to contractors may be creating a conflict of interest.

**Recommendation.** DOE project management should be staffed to the level needed to ensure that the government's interests are protected. DOE should assess whether it has enough project management personnel to properly discharge its ownership role or whether DOE understaffing in project management is permitting contractors to take on responsibilities and functions that should be reserved for the government's representatives. To do this, DOE will have to define the roles and responsibilities of federal project managers and then assess the number of project managers needed to carry out these responsibilities. The roles and responsibilities of the contractors' project managers vis-à-vis the federal project managers should also be clarified.

**Recommendation.** DOE should develop a vision for what project management in the department should become, and then hire, train, and promote personnel specifically to staff and fulfill this vision.

**Recommendation.** Concurrent with the DOE staffing assessment, DOE should also assess the project management staffing of its (M&O and M&I) contractors in terms of both quantity and quality (knowledge, background, and experience). It would be desirable to know if contractors, perhaps because of the declining competition for DOE projects, are not assigning their best managers to DOE projects.

**Recommendation.** DOE should estimate its future requirements for project management and other project support personnel and develop a plan to address recruitment, turnover, and retention in the future. Hiring personnel with experience in preproject planning, cost estimating, risk management, EVMS, team facilitation, and other critical skills can be a means of meeting some of those needs in the near term.

**Finding.** The committee perceives a need for improved utilization of existing and incoming project management personnel. This need can be fulfilled through training and career development and by facilitating the movement of personnel across organizational lines. Executing the PMCDP as a DOE-wide program will go a long way toward overcoming present training deficiencies. However, a long-term commitment to funding implementation of the PMCDP is critical.

**Recommendation.** The projected annual tuition expenditure for training and development of $1.5 million is considered adequate for the immediate concentrated need. Every effort should be made to allocate this amount centrally based on a DOE-wide decision, especially in the first few years, to assure implementation of the PMCDP throughout the organization. In the interim, the DOE field and project offices should continue to meet immediate needs with their own training programs.

**Recommendation.** In a previous report, an NRC committee recommended that DOE should "develop and maintain a cadre of professional certified project managers who would be assigned to manage DOE projects for all program offices" (NRC, 1999, p. 77). Since it is clear that DOE does not intend to implement this recommendation, the committee recommends that DOE treat qualified project management personnel as a shared resource and facilitate their movement to assignments across the organization as the needs arise. OECM, in conjunction with the operation of the PMCDP, should maintain an inventory of all project managers throughout the DOE complex, along with their experience and capabilities, and make this inventory available to all DOE programs as they staff their projects.

# AWARDS

## 2002

**Finding.** DOE has executed several recent projects successfully and on or ahead of budget and schedule, as indicated by its 2002 project management award program.

**Finding.** While all projects considered for 2002 awards were initiated prior to the publication of DOE Order O 413.3, the principles and procedures required by the order and outlined in this and prior committee reports were important factors in successful completion.

**Finding.** Lessons learned from briefings by award recipients have application to project personnel who did not attend the Project Management Workshop.

**Recommendation.** Copies of briefings by the 2002 award recipients should be distributed to all field offices that have project personnel.

**Finding.** [The National Nuclear Security Administration] NNSA and EM were the only program offices that participated in the 2002 project management awards program. Other DOE offices that execute projects had no nominations.

**Recommendation.** DOE should determine why the other program components did not participate in the awards program. DOE should encourage full participation in the future.

**Finding.** The Project Management Workshop is a step forward in recognizing exemplary projects and project managers and in building a sense of professionalism among project personnel.

**Recommendation.** DOE should continue and even expand this workshop in future years.

## PROJECT REVIEWS

## 1999

**Finding.** Independent project reviews are essential tools for assessing the quality of project management and transferring lessons learned from project to project.

**Finding.** External independent reviews of 26 major projects are under way to assess their technical scope, costs, and schedules. The reviews so far have documented notable deficiencies in project performance verifying the committee's conclusion that DOE's project management has not improved and that its problems are ongoing. However, DOE has yet to formalize and institutionalize a process to ensure that the recommendations from these reviews are implemented.

**Finding.** Various DOE program offices are also developing the capability of conducting internal independent project reviews.

**Recommendation.** DOE should formalize and institutionalize procedures for continuing independent, nonadvocate reviews, as recommended in the Phase I report of the National Research Council to ensure that the findings and recommendations of those reviews are implemented. DOE should ensure that reviewers are truly independent and have no conflicts of interest.

**Recommendation.** All programs that have projects with total estimated costs of more than $20 million should conduct internal reviews, provided that the value of the reviews would be equal to or greater than the costs of conducting them. Deciding if an internal review is justified for a given project should be the joint responsibility of program management and the project management organization. The decision should be based on past experience with similar projects, the estimated cost of the project, and the uncertainty associated with the project. Internal reviews are expensive and take up the time of valuable people, so they should not be undertaken lightly. However, under the present circumstances, the committee believes that more internal reviews would be justified. The project management organization should manage these reviews for the director or assistant secretary of the cognizant program office. The results of these reviews should be taken by the program office to the Energy Systems Acquisition Advisory Board (ESAAB) and used as a basis for the decision whether to continue the project.

## 2001

**Finding.** The evidence available to the committee indicates that the [external independent review] EIR program continues to identify significant management issues in the projects reviewed and in DOE's operation in general. Absent substantial evidence of improvement in DOE project management and project performance, the EIR program needs to be continued.

**Recommendation.** DOE, through OECM, should establish performance metrics for the EIR program that identify trends and opportunities for improving project management performance.

**Recommendation.** The EIR program should continue in its present form under

OECM direction until there is clear evidence of improvement in DOE project management and project performance.

**Finding.** DOE would benefit from a department-wide procedure governing external and internal independent reviews. Consistent procedures would increase the pool of qualified reviewers, expedite the review and report process, and enable an automated system for tracking deficiencies and corrective actions.

**Recommendation.** DOE should expedite the issuance of the Independent Review Procedure drafted by OECM.

**Finding.** A more thorough review analysis for defining mission need and setting a preliminary baseline range during the front-end project planning phase would give the decision makers more useful information.

**Recommendation.** DOE should expand the use of [internal project reviews] IPRs for the CD-0 decision and should require an [independent cost review] ICR prior to CD-1.

**Finding.** There is some concern that mandating formalized reviews for projects costing between $5 million and $20 million TPC may be dedicating manpower and money beyond the point of significant value added. It also may be distracting project personnel from doing the project and diverting DOE's project management resources from larger, more complex projects.

**Recommendation.** DOE should reevaluate the benefits gained from mandating reviews for projects costing between $5 million and $20 million TPC. The OECM should establish guidelines to evaluate the cost-effectiveness of review. At a minimum, the $5 million threshold should be based on TEC and provide for significant tailoring of the review process.

**Finding.** The EM Project Definition Rating Index (EM-PDRI) analyzes the readiness of a project by rating it on a numerical basis. It allows making judgments based on a multitude of rating factors, including risk, but users will need training and experience with the index in order to achieve uniformity of application and confidence in the results.

**Recommendation.** DOE should explore the potential application to other programs of the PDRI approach adopted by EM.

# ACQUISITION AND CONTRACTING

## 1999

**Finding.** Traditional DOE contracting mechanisms, such as cost-plus-award-fee and management and operations (M&O) arrangements, are not always optimal for DOE's complex mission. These approaches are being replaced with more effective approaches based on objective performance incentives, but change has been slow.

**Finding.** DOE's long history of hiring contractors to manage and operate its sites on the basis of cost-plus-award-fee contracts has created a culture in which neither DOE nor its contractors is sufficiently accountable for cost and schedule performance.

**Finding.** DOE does not use effective performance-based incentives and does not have standard methods for measuring project performance.

**Finding.** DOE does not effectively match project requirements and contracting methods. Mismatching often results in cost and schedule overruns.

**Finding.** The numbers of bidders on major DOE contracts has been declining and in some cases have not elicited truly competitive bids. This may indicate that projects are not being appropriately defined and packaged and that the disincentives to bid often outweigh the incentives.

**Recommendation.** DOE should strengthen its commitment to contract reform focusing on the assessment and quantification of project uncertainties, the selection of the appropriate contract type and scope for each job, and increased use of performance-based incentive fees rather than award fees to meet defined project cost and schedule goals. A comprehensive risk analysis should be conducted before deciding whether to issue fixed-price contracts for work that involves a high level of uncertainty (such as new technology or incomplete characterization). Specific contract scopes and terms should be negotiated to define both DOE and contractor responsibilities to prevent cost overruns. Clear, written roles, authorities, and responsibilities should be established for DOE headquarters, field elements, contractors, and subcontractors for each contract. Guidelines should be provided for the appropriate times in the project for the selection of contractors.

**Recommendation.** DOE should develop written guidelines for structuring and administering performance-based contracts. The guidelines should address, but need not be limited to, the following topics: the development of the statement of work; the allocation of risks to whomever would be most effective at controlling the risks (either DOE or the contractor); the development of performance measures and incentives; the selection of the contracting mechanism; the selection of the contractor; the administration of the contract; and the implications of federal and DOE acquisition regulations. DOE should train its employees in the roles and responsibilities of a performance-based culture and then hold both employees and contractors accountable for meeting these requirements.

**Recommendation.** DOE should provide financial rewards for outstanding contractor performance to attract bids from the best contractors. A DOE-wide policy should be developed that provides fiscal rewards for contractors who meet or exceed schedule, cost, and scope performance targets. Contractor fees should be based on contractor performance.

**Recommendation.** DOE employees and contractor employees essential to projects should be trained in acquisition and contract reform. The training of source selection officials and members of source evaluation boards should be expedited; a minimum level of training should be a prerequisite.

## 2001

**Finding.** The extent of training and use of performance-based contracting (PBC) in DOE contracting efforts is unclear. There is no DOE-wide database that shows the extent of use of PBC or the number of staff trained in PBC techniques.

**Recommendation.** The committee reaffirms the recommendations made in previous reports (NRC, 1999, 2001) on using PBC and encourages OECM to play a lead role in supporting this practice. OECM should work closely with the Office of Procurement and Assistance Management to see that PBC training is provided as part of the career development process for project management personnel and just-in-time training for the [integrated project team] IPT. In the near term, OECM should bring on board a cadre of experts, skilled in performance-based contracting, to provide technical assistance to IPTs responsible for new major system initiatives.

**Finding.** The draft *Program and Project Management* (PPM) manual and draft *Project Management Practices* (PMP) developed by the OECM fail to address PBC adequately.

**Recommendation.** The detailed descriptions of PBC alternatives and their application to DOE projects should be included in the revised PMP and PPM.

**Finding.** There have been continuing efforts on the part of DOE to move toward a more effective use of PBC methods and to support these efforts.

**Recommendation.** Contract approaches should be tailored to use fixed-price and performance-based methods where practicable to assist the DOE to get the most cost-effective results and to stimulate competition. In addition, the department should continue to explore other innovative commercial contracting approaches to meet its needs.

## 2002

**Finding.** The committee believes that the August 2002 draft PPM provides a good framework for addressing acquisition strategy issues and offers a useful model for DOE project managers to follow in preparing and planning their efforts. However, it finds that the acquisition strategy documents being reviewed by OECM are of mixed quality and believes that this indicates a need for more training and development of additional reference documents. The iterative process of review and correction will also improve the overall quality of planning documents over time. The most recent draft PPM now provides clear and consistent guidance on what needs to be addressed in each draft acquisition strategy and, as such, should significantly increase the quality of the documents submitted for review.

**Recommendation.** The committee recommends that senior management continue to require project teams to focus on up-front acquisition planning and that it continue to use the approval process for ensuring compliance and consistency. DOE management should return documents that do not meet management expectations and should follow up by asking why these inadequacies were not fixed at lower levels. Project teams should be trained in developing effective acquisition strategies. DOE leadership should also continue to focus on competition to obtain a range of innovative approaches from a variety of contractors to meet its management, operating, and development needs.

## PERFORMANCE-BASED CONTRACTING

## 2002

**Finding:** For large cost-reimbursement contracts, many factors compromise the ability of the government to use purely objective measures for assessing performance. Moreover, federal agencies are comfortable using a more traditional cost reimbursement award fee approach, in which the award fee is at the discretion of the federal project manager.

**Recommendation:** A significant amount of up-front planning by the IPT is needed to specify outcomes and identify those aspects of an overall project for which a contractor can effectively be held accountable. The committee reiterates its recommendation that training in performance-based contracting methods be provided to IPT members. In addition, DOE should collect best practices information on the use of performance-based contracting in DOE contracts and identify those activities most suitable for use of these metrics.

**Finding.** DOE is reassessing its cleanup efforts, giving them a new focus on cost-effective and rapid closure of sites, and setting up incentives that can best achieve that goal.

**Recommendation.** DOE should reassess its use of incentives in existing contracts to ensure that they focus on closure and that interim goals are effective in driving this overall objective.

## ORGANIZATIONAL STRUCTURE, RESPONSIBILITY, AND ACCOUNTABILITY

### 1999

**Finding.** DOE's organizational structure makes it much more difficult to carry out projects than in comparable private and public sector organizations. Successful corporations and agencies responsible for major projects arrange their organizations to provide focused and consistent management attention to projects.

**Finding.** Too many people in DOE act as if they were project managers for the same project, and too many organizations and individuals outside the official project organizations and lines of accountability can affect project performance.

**Finding.** Compliance with DOE's policy requiring the establishment of performance agreements and self-assessments from the field has been limited and slow.

**Recommendation.** To improve its project management performance, DOE should establish an office of project management on a level equal to or higher than the level of the offices of assistant secretaries. Department-wide project management functions should be assigned to the project management office, and the director of this office should have the authority and the resources to set and enforce reporting requirements for all projects. Other responsibilities, such as property and asset management, should be assigned to existing DOE headquarters offices. To be successful, the office of project management must have the full and continuing support of the secretary, the under secretary, the deputy secretary, and of all of the program offices and field offices as a top-down management initiative.

## 2001

**Finding.** DOE continues to rely excessively on contractors for project justification and definition of scope. There are some large projects in which DOE is not effectively executing its role of owner with respect to the oversight and management of contracts and contractors.

**Recommendation.** DOE should develop its position as an effective owner of projects and should assure that federal project managers are trained and qualified owner's representatives, capable of dealing effectively with contractors.

**Finding.** The combination of the OECM and the PMSOs in the three major PSOs addresses many of the issues raised in the Phase II report but not all. This organizational structure is probably workable, but it does not fully address the department-wide issues of consistency, discipline, and excellence in project management that the Phase II committee felt were essential.

**Recommendation.** The roles and responsibilities of the OECM should be strengthened, as set forth in the Phase II report, and the OECM should be budgeted, staffed, and empowered to become the center of excellence in project management and the coordinator for project manager training and development and for oversight and approval of all capital projects in DOE.

## 2002

**Findings.** The forum held in November 2001 provided examples of the points made in the committee's previous reports about how industry fulfills its role as owner in planning and managing projects. In subsequent meetings with DOE and DOE contractor personnel the committee saw evidence of increased emphasis on front-end planning and a clearer understanding of DOE's role as owner. Recent policy memoranda that emphasize acquisition planning are encouraging.

**Recommendation.** The committee believes that in order for DOE to be an effective owner of capital acquisition projects it should:

- Consider capital projects critical to organizational success.
- Require senior management involvement in project decision making, usually at the $5 million dollar and higher level.
- Have a detailed and well-recognized internal front-end planning process.
- Capture metrics on planning effort and project performance.
- Require owner involvement and leadership in front-end planning.
- Ensure that projects support DOE's mission and are consistent with DOE's strategic plan.

**Recommendation.** DOE should periodically benchmark its performance in project planning and control processes and compare it with the performance of industry leaders to ensure that it is consistently utilizing the best practices.

**Recommendation.** Senior managers in each program secretarial organization (PSO) in DOE should develop a complete definition of the roles and responsibilities of project managers.

**Recommendation.** Senior managers should continue to emphasize the importance of improving the project management processes and procedures to assure long-term improvement throughout the organization.

## PROJECT MANAGEMENT CULTURE

### 2002

**Finding.** DOE personnel and contractors generally support the need for a comprehensive project management system but prefer a system with fewer requirements for upper management oversight.

**Recommendation.** DOE should resist efforts to reduce up-front planning requirements and to lower the level of authority at which critical decisions are approved. DOE should apply persistent pressure to ensure that the right projects are picked for execution and that they are planned and executed according to established policies and procedures. Procedures should continue to include a process for tailoring requirements to the size and complexity of projects.

**Recommendation.** DOE should assess its culture and subcultures and develop strategies to bring about organization alignment on core project management principles at all levels of the organization.

## REFERENCES

NRC (National Research Council). 1999. Improving Project Management in the Department of Energy. Washington, D.C.: National Academy Press.
NRC. 2001. Progress in Improving Project Management at the Department of Energy, 2001 Assessment. Washington, D.C.: National Academy Press.
NRC. 2003. Progress in Improving Project Management at the Department of Energy, 2002 Assessment. Washington, D.C.: The National Academies Press.

# Delegation of Authority as Acquisition Executive

**The Under Secretary of Energy**
Washington, DC 20585

April 11, 2003

MEMORANDUM FOR RAYMOND L. ORBACH
DIRECTOR, OFFICE OF SCIENCE

FROM:                          ROBERT G. CARD
                               UNDER SECRETARY
                               ENERGY, SCIENCE AND ENVIRONMENT,
                               AND ACQUISITION EXECUTIVE

SUBJECT:                       Delegation of Authority as Acquisition Executive

The Department instituted DOE Order 413.3, Program and Project Management for the
Acquisition of Capital Assets, on October 13, 2000 and a companion project management
manual on March 28, 2003. The Order and Manual provide direction for the acquisition
of capital assets. They also provide for the delegation of authority as Acquisition
Executive as the senior manager responsible and accountable for all acquisition of capital
assets. As such, I hereby delegate to you the Acquisition Executive (AE) authority for
the acquisition of capital assets that have a total cost of $400M and below. You may
further delegate this AE authority as provided in the Order and Manual. I am also
delegating to you the authority to approve performance baseline changes that are below
the Secretarial Acquisition Execution approval level.

This authority is subject to the condition that you must have your project management
system approved by me within 12 months from the date of this delegation or the
delegation will be withdrawn.

Furthermore, any appropriate and legitimately appointed Acting Assistant Secretary may
exercise this AE authority during periods when you are not available due to extended
travel.

This delegation will remain in force until superceded or rescinded. Implementing
guidance and assistance is available from James A. Rispoli, Director, Office of
Engineering and Construction Management (ME-90).

# Summary of External Independent Reviews for National Nuclear Security Administration, Office of Environmental Management, and Office of Science Projects, FY 2001 Through FY 2003, Covering Only Baseline Validation Prior to Critical Decision 2

This summary was prepared by the DOE Office of Engineering and Construction Management and covers 19 projects with total estimated costs of under $100 million, equally apportioned between the Office of Science (SC), the Office of Environmental Management (EM), and the National Nuclear Security Administration (NNSA).

### NATIONAL NUCLEAR SECURITY ADMINISTRATION PROJECTS

Project:    Terascale Simulation Facility
Site:       Sandia National Laboratories
ID:         00-D-103
TPC:        $95.3 million
EIR Date:   August 2001
Key Findings:
  • Resource-loaded schedule was not developed.
  • Fiber-optic system was not included in the cost estimate.

Project:    Test Capabilities Revitalization Phase I
Site:       Sandia National Laboratories
ID:         02-D-103-01
TPC:        $47.3 million
EIR Date:   December 2002

Key Findings:
- Project has three tiers of equipment. Tiers 2 and 3 appeared to be desirable items that will be purchased if money is available at the end of the project. There was no documented basis of what was in the baseline.
- No documented connection exists between the results of the risk analysis workshops and the contingency figures the project team actually applied to the project estimate.
- The backup support for the demolition and disposal estimate does not match the baseline estimate and is not detailed enough to support the $1 million estimated for this activity.

Project:     Gas Transfer Capacity Expansion project
Site:        Kansas City
ID:          03-D-121
TPC:         $31.3 million
EIR Date:    November 2002
Key Findings:
- Key assumptions for cost and schedule estimates were not documented.
- Assumptions for overhead and profit markups were unreasonable.

Project:     Joint Computational Engineering Laboratory
Site:        Sandia National Laboratories
ID:          00-D-107
TPC:         $30.8 million
EIR Date:    September 2001
Key Findings:
- Funding level had not been established consistent with project costs.
- Project did not have a current and sound estimate of TEC and TPC.

Project:     Sensitive Compartmented Information Facility
Site:        Lawrence Livermore National Laboratory
ID:          00-D-800
TPC:         $25.1 million
EIR Date:    July 2001
Key Findings:
- Project schedule was not complete.
- A critical path analysis was not performed. Key equipment requiring long-lead procurement was not identified.
- Review all drawings and calculations and ensure that the design criteria requirements are met.
- Hazards analysis, including fire hazards, not complete.

Project: Bus Upgrades
Site: Nevada Test Site
ID: 02-D-107
TPC: $16.7 million
EIR Date: September 2002
Key Findings:
- Project schedule does not have sufficient detail for performance baseline.
- Cost baseline is not in agreement with funding profile.
- Drawings were not consistent with project baseline. For example, retention pond shown in the drawings was not in the baseline.

Project: Office Building Replacement
Site: Los Alamos National Laboratory
ID: 01-D-704
TPC: $10.5 million
EIR Date: May 2001
Key Findings:
- Schedule does not include all significant activities and is not integrated.
- Requirements are not clearly defined (i.e., document has conflicting statements). For example, section 15101 states that a complete chilled water system shall be provided. Performance specification section 15670 states that the chilled water system is an option based on the results of a life-cycle cost analysis.
- Project has not been sufficiently planned to implement the federal government's goals for efficient energy management.

## OFFICE OF ENVIRONMENTAL MANAGEMENT PROJECTS

Project: Glovebox Excavation Method Project—Pit 9
Site: Idaho National Engineering and Environmental Laboratory
ID: ID-ER-106
TPC: $79.6 million
EIR Date: July 2002
Key Findings:
- Funding requirements by FY could not be verified due to a lack of crosswalk from estimate to schedule.
- There was no current approved baseline schedule that reflected the current plan.
- Some technical issues of risk had not been identified in the risk assessment, there was no crosswalk from the project risk assessment to the project action item list, and no quantitative schedule risk analysis had been done.

- System design criteria documents contained some outdated and/or incomplete information. In some specific cases, they did not reflect current project requirements for use in detailed design.
- No systematic consideration or application of value engineering to address life-cycle costs.

Project: Ashtabula Closure Project
Site: Ohio
ID: OH-AB-0030
TPC: $43.4 million
EIR Date: June 2003
Key Findings:

- The development of the site's long-term stewardship (LTS) plan is not captured in the work breakdown structure.
- Several contract issues are not adequately addressed.
- The risk management process is based on a qualitative analysis and provides no definitive information on the potential magnitude of the cost and schedule impacts associated with the risks. Government furnished services and items (GFS&I) deliverables and some contractor risk items were not included in the risk analysis and plan. No contingency or reserve for changes to contractor's scope or other risks.
- The sanitary batch reactor is scheduled for demolition in September 2003 and there is no alternative plan in place to handle sanitary wastes in support of remediation activities scheduled through May 2005.
- There is no sitewide groundwater remediation strategy and no basis for the contractor's commitment for a pump-and-treat remediation system.
- There is no plan for a reduction in staffing tied to any environmental safety and health (ES&H) milestones to reflect the project winding down, nor does Human Resources appear to be intimately involved.
- The cleanup level is well below the typical cleanup levels found at other sites. Significant cost savings/avoidance could be achieved if higher levels are accepted for this site.
- The contractor does not have a resource-loaded integrated schedule. It uses a combination of Microframe to develop earned value, Primavera P3 for schedules, and Cost Point to collect costs. The systems do not use common milestones.
- The total project cost is inconsistent with the funds analysis.
- Statements have been made regarding the acceleration of the remediation schedule with added funding, but there is no supporting documentation.
- Schedule and cost estimate were developed without an allowance for overtime.

Project:    Plutonium Packaging and Stabilization (Pu P&S) Project
Site:       Savannah River Site
ID:         01-D-414/02-D-420
TPC:        $25.5 million
EIR Date:   April 2002
Key Findings:
- The manner in which the escalation was calculated did not provide an accurate estimate of project costs.
- Project contingency appeared to be high for this project at its stage of design development.
- The performance criteria and requirements regarding functional capabilities were not complete.
- The draft project execution plan (PEP) was incomplete and not current.
- The project change control thresholds did not correspond to O 413.3.
- Some major equipment items included in the baseline were not included in the Acquisition Strategy.
- There was no crosswalk to assure that project risk implications of open technical issues had been addressed.

Project:    INEEL CERCLA Disposal Facility (ICDF)
Site:       Idaho National Engineering and Environmental Laboratory
ID:         ICDF
TPC:        $46.9 million
EIR Date:   March 2002
Key Findings:
- The current budget/funding requirements by FY could not be verified.
- The most current budget guidance for EM projects was not used in setting escalation rates for construction and life-cycle costs.
- Schedule activities were not man-hour loaded to assure adequate duration of activities and that activities are appropriately sequenced.
- The PEP was not final and did not conform to O 413.3.
- The baseline change control process did not conform to O 413.3.
- Not all risks to the project had been identified. Cost and schedule implications of identified risks had not been quantitatively evaluated. Progress of risk reduction and mitigation activities was not tracked.
- No time frame for accomplishing risk reduction/mitigation actions was provided.

Project:    Western Sector Dynamic Underground Stripping Project
Site:       Savannah River Site
ID:         CA-1707
TPC:        $18.6 million
EIR Date:   October 2002

Key Findings:
- Several of the larger work activities had a 32 percent markup. Some project-specific factors support a higher-than-normal overhead and profit markup; this one is too high for large construction activities.
- The well drilling and installation estimate is based on a few assumptions and a generalized work scope.
- Escalation was not applied to a major portion of the construction estimate for FY 2003.
- The 10 percent construction contingency appears to underestimate the risk associated with the installation of the dynamic underground stripping remediation system.
- The PEP did not clearly identify the schedule baseline or adequately describe the technical baseline.
- The project summary schedule does not include adequate detail.
- The schedule of contractor activities was not adequately detailed for negotiation.
- No signed mission-need approval letter from DOE headquarters had ever been issued.

Project:    Miamisburg Environmental Management Closure Project, Potential Release Site (PRS) 66 Proposed Remedial Action
Site:       Ohio
ID:         OH-MB
TPC:        $11.7 million
EIR Date:   March 2003
Key Findings:
- No significant issues.
- The scopes of the PRS 66 remedial action plans are complete and well defined. They are based upon a comprehensive characterization leading to a good definition of the location of waste and the amount of soil needed to be removed.
- The uncertainties with the remediation effort are defined and are generally provided for in planning for the project.

Project:    INTEC Cathodic Protection System Expansion Project
Site:       Idaho National Engineering and Environmental Laboratory
ID:         02-D-402
TPC:        $6.7 million
EIR Date:   September 2001
Key Findings:
- The Construction Project Data Sheet does not reflect the current cost estimate.
- The intended work scope was not clearly outlined in the Title I design documents.

- The work scope definition and design criteria are not clearly documented.
- There is no documentation of a formal design review having been conducted when the Title I design was completed.
- The project management estimate is higher than recommended in DOE cost guidelines.
- The durations used to calculate the escalation are not consistent with the project schedule, and the assumed average escalation of 2.9 percent is higher than DOE published rates.
- The cost estimate and the scope of work are conservative in several areas and have a built-in contingency in several areas.
- The support from radiation contamination technicians and the disposal of mixed and radioactive wastes are included in other project costs (OPC) but should be in the total estimated cost (TEC).

## OFFICE OF SCIENCE PROJECTS

Project:    Mechanical and Control Systems Upgrade—Phase 1
Site:       Argonne National Laboratory-East
ID:         MEL-001-17
TPC:        $9.1 million
EIR Date:   September 2002
Key Findings:
- The project does not have a fully integrated resource and cost loaded critical path method schedule.
- The cost for risk mitigation contingency shown in the project cost estimate was not derived from the application of the risk model.
- No commissioning plan exists to ensure that the replacement systems, equipment, and components are ultimately capable of being operated and maintained according to the laboratory's operational needs.
- No quality assurance plan that both assigns roles and responsibilities and defines procedures specific to this project has been developed.
- No construction inspection and acceptance testing plan specific to the systems, equipment, and components to be replaced in this project has been developed.

Project:    Center for Nanophase Materials Sciences
Site:       Oak Ridge National Laboratory
ID:         03-R-312
TPC:        $65 million
EIR Date:   August 2002
Key Findings:
- Escalation included in the cost estimate did not use the best process and was based on FY 2003 rather than FY 2004 guidance.

- Life-cycle costs have not been developed for the total project.
- The current schedule has not been cost and resource loaded.
- The current summary-level project schedule is incomplete because it does not include some major project activities—for example, Title III, project management, design support, OPC activities, and commissioning.
- The ES&H plan at the University of Tennessee (UT)-Battelle (O&M contractor) level has not been formally adopted.
- Analysis supporting design decisions was not well documented.
- There is no system in place to ensure that the building systems are designed, installed, functionally tested, and capable of being operated and maintained according to UT-Battelle's operational needs and that the building systems will meet UT-Battelle's needs.
- The PEP has not been finalized.
- An overarching quality assurance plan at the UT-Battelle level, which covers both equipment for the center and construction of the center, has not been established.
- The level of detail and specificity in the risk assessment should be increased.
- A system to review the constructibility, buildability, and bidability of the design has not been put in place.

Project:      RUN IIB D-Zero and CDF Detectors
Site:         Fermi National Accelerator Laboratory
ID:           SC-1/2
TPC:          $59 million
EIR Date:  December 2002
Key Findings:

- The PEP for the project is incomplete. Value engineering, quality assurance, and risk management are neither addressed nor referenced in the PEP.
- Project-specific configuration management and control process have not been developed, as required by O 413.3. Further, no laboratory configuration management and control policy or procedure is cited in the project management plan (PMP).
- There is no description or reference in the PEP, or in the PMP, to the flow-down of requirements and processes for quality assurance and quality control to specifics of design, fabrication, procurement, or establishment and maintenance of document approval and authenticity.

Project:      Research Support Center
Site:         Oak Ridge National Laboratory
ID:           MEL-001-25
TPC:          $16.3 million
EIR Date:  August 2002

Key Findings:
- The Title I (CD-2) cost baseline has not been formally established, and various versions of the baseline exist throughout the project documentation, negating the usefulness of a controlled baseline estimate.
- Costs have been shifted to other projects for some items and some functionality has been eliminated from the center. The driving force behind this effort is to keep the total estimated cost at $16.1 million.
- The project integration estimate lacks backup support, activities are not defined, and a cost basis is not provided.
- The estimate contains no costs for postconstruction activities such as commissioning, development of maintenance and safety plans, and management of these activities.
- The baseline schedule does not have sufficient detail to support CD-2.
- The level of detail for an $11 million construction subcontract is inadequate to justify the activity's duration. No detail was available to justify further proposed reduction in the duration for this activity from 22 months to 16 months.
- A cost-loaded schedule is not available, as required by O 413.3.
- The baseline control and project control milestones included in the PEP use a timeline graduated in FY quarters. This timeline scale is too general for the change control thresholds proposed in PEP.
- UT-Battelle indicated it intends to purchase standard equipment for the center as a separate item. No line item on the schedule shows these procurements or their coordination with the facility construction.
- The schedule does not contain an identified critical path.
- The scope baseline in PEP Section 5.1 states that the center will be a 50,000-square-foot building. The reviewers were advised that this is no longer the case, and a 53,500-square-foot building is currently planned.
- Project documentation provided to the review team does not consistently reflect the current scope.
- The project does not have a clearly defined integrated project team (IPT).
- The PEP approved in May 2002 is now outdated and in need of extensive revision.
- The risk assessment plan identifies only three risks, leaving many unaddressed—for example, coordination perils resulting from the concurrent general plant projects, state and private sector projects occurring in the same vicinity, and the likelihood of funding shortfalls. The risk mitigation strategies lack specificity and depth.
- There is no documented quality assurance program, and the review team was informed that a documented program for the project is not intended. The organization charts and descriptions of project responsibilities do not show an individual assigned to quality assurance.

- The East Research Campus area will become the scene of increasing construction activity as simultaneous projects conducted by different parties for different owners compete for utilities, space, and resources. The PEP does not include any description of a coordination effort or organization necessary to manage this complex project area.
- The undated acquisition execution plan (AEP) lacks specificity, includes contradictory statements, and is inconsistent with other project documents.
- The cost and schedule baselines cannot be determined from the documentation. The documents contain no consistent "official" set of cost and schedule figures.
- The PEP does not adequately describe a baseline change control board function.
- The configuration management process described in the PEP does not sufficiently describe the process used for configuration management, the coordination requirements with others, or the level of effort planned for this function.
- The project cost estimates include wildly fluctuating contingency amounts as the project team endeavors to maintain the TEC.

Project:     Laboratory Systems Upgrades
Site:        Pacific Northwest National Laboratory
ID:          MEL-001-18
TEC:         $9.4 million
EIR Date:    May 2002
Key Findings:

- The install and test control system costs are higher than normal, and Johnson Controls, Incorporated, the contractor designing the system and making preliminary cost estimates, is being considered for a sole-source procurement.
- The productivity rates assumed for a number of construction activities appear overly conservative.
- The contingency level in the project team's estimate is higher than the level recommended in DOE guidance and does not appear justified.
- The project team has not sufficiently defined milestones included in the schedule baseline to manage the project in accordance with the project baseline change control thresholds and DOE guidance.
- Several key areas of the preliminary design were missing outline specifications, equipment specifications, general notes, and keyed notes.
- The documentation did not meet the minimum requirements for receiving CD-2 approval.
- There is no risk management plan. The Contingency and Risks section of the PEP does not sufficiently identify and quantify risks or define mitigation strategies.

# Correspondence Between Dr. Orbach and Dr. Alberts Regarding the Committee's 2002 Assessment Report

**Department of Energy**
Office of Science
Washington, DC 20585

June 16, 2003

Office of the Director

Dr. Bruce M. Alberts
President
National Academy of Sciences
2101 Constitution Avenue, N.W.
Washington, DC 20418

Re:   NRC Report: *Progress in Improving Project Management at the Department of
        Energy: 2002 Assessment*

Dear Bruce:

In performing my duties I rely heavily on advice and recommendations from many
external independent advisory groups. I view constructive criticism as a healthy and
useful mechanism to enhance the performance of SC programs and projects.

Congress asked the NRC to assess the Department of Energy's (DOE) progress in
improving its project management practices. This effort, initiated in FY 1999, is nearly
complete. The NRC's Committee charged with performing the assessment recently
issued *Progress in Improving Project Management at the Department of Energy: 2002
Assessment*, the second of three annual assessments.

On June 11, 2003, I met with the Committee Chair and several other Committee
members. I appreciated the opportunity to voice my support for improvements in DOE's
project management practices. SC strongly supports improving project management
within the Department of Energy and appreciates the Committee's efforts and insights. I
place a high priority on executing the right projects the right way; ensuring the presence
of qualified and experienced Project Directors; and institutionalizing a rigorous yet
appropriately flexible project management system.

Unfortunately, the Committee included two categorical comments, which, in my view,
detract from their careful and helpful detailed analysis. They are:

> 1)  Fourth paragraph, page 13:

> "The NRC report *Improving Project Management in the Department of Energy*
> (NRC, 1999) stated that cost increases are often distorted by DOE's tendency to
> consider project scope as a contingency. This situation still prevails, particularly
> in the Office of Science (SC), which tends to use a design-to-budget approach."

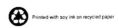

Printed with soy ink on recycled paper

2) Second paragraph, page 38:

"The SC mission, on the other hand, is developed by scientists in the DOE national laboratories, which are run by university contractors and are naturally resistant to direction from Washington."

The first assessment, "a design-to-budget approach," suggests that SC does not maintain project scope. No example is given or quoted that would give credence to this categorical conclusion, so it is hard for me to respond, but I believe it is based on erroneous information given to the Committee regarding the Spallation Neutron Source (SNS) currently under construction. That project scope did increase from the original project baseline, but without increasing the cost of the project—even though it is a very complex scientific instrument costing $1.4 billion for construction. I am disappointed that the Committee continued to support its categorical assessment of SC even when presented with a full explanation of the SNS scope.

The second assessment, "university contractors...naturally resistant to direction from Washington," unfairly characterizes the hard working committed leadership of SC science laboratories. In my tenure as Director, I have experienced only the closest and warmest relationship with university contractors. They have been extraordinarily responsive to direction from Washington. Again, I am disappointed that the Committee maintained its categorical assessment even when presented with the Washington view (mine).

I believe the Committee did an excellent job in its analysis of Department of Energy's project management. As I stated above, I welcome constructive criticism, well informed and targeted. However, the credibility of the Committee's report has been seriously damaged by their inaccurate categorical statements, and has unfairly smeared my (SC) organization.

I shall continue to support the efforts of the Committee for Oversight and Assessment of U.S. Department of Energy Project Management. However, in the future, I ask the Committee to be more sensitive to the impact inaccuracies can have on the credibility of its own assessment and the affected organizations.

Sincerely,

Raymond L. Orbach
Director

cc:   R. Card, Under Secretary
J. Powers, DOE/ME-20
J. Rispoli, DOE/ME-20
D. Lehman, DOE/SC-81
K. Reinschmidt, NAS
M. Cohen, NAS

### NATIONAL ACADEMY OF SCIENCES
*THE NATIONAL ACADEMIES*

Office of the President

July 15, 2003

Dr. Raymond L. Orbach
Director
Department of Energy
Office of Science
1000 Independence Avenue, SW
Washington, D.C. 20585

Dear Ray:

Thanks for your letter of June 16 regarding the recent NRC report, "Progress in Improving Project Management at the Department of Energy: 2002 Assessment." I appreciate your candid feedback voicing "support for improvements in DOE's project management practices", while expressing concern about several of the committee's findings. As you are no doubt aware, this committee has a long history of providing advice on project management of major activities at DOE. It has been faced less often with reconciling changing mission responsibilities with the "nuts and bolts" of project management per se, their principal charge in this case.

In particular, the report conveys the consensus opinion of the committee based on its review of numerous documents and discussion with DOE personnel and DOE contractors in Washington, D.C. and at various sites around the nation. The committee's observations and opinions might have been modified somewhat, however, if it had the opportunity to focus more on the relationship between the scientific mission, with its frequent changes, and the much more tactical aspects of project management. The Office of Science is perhaps an area in DOE where these responsibilities are most intertwined and obtaining a complete picture is difficult. In the next round, we are seriously considering adding some committee members who have had experience with the nature and operations of this department.

We appreciate your continued support for the efforts of the committee. Their next report will assess progress in improving project management in the DOE since 2000 and the probability that the department can sustain its achievements and continue to improve. The committee certainly benefited from the discussion with you on June 11 and would be very pleased to consider any additional information that you think appropriate as it begins it next assessment.

With best regards,

Bruce Alberts
President

**THE NATIONAL ACADEMIES**
*Advisers to the Nation on Science, Engineering, and Medicine*

2101 Constitution Avenue, NW
Washington, DC 20418

*Mailing address:*
500 Fifth Street, NW
Washington, DC 20001